ECONOMIC RESTRUCTURING AND POLITICAL RESPONSE

Volume 34, URBAN AFFAIRS ANNUAL REVIEWS

ECONOMIC RESTRUCTURING AND POLITICAL RESPONSE

Edited by
Robert A. Beauregard

Volume 34, URBAN AFFAIRS ANNUAL REVIEWS

SAGE PUBLICATIONS
The Publishers of Professional Social Science
Newbury Park London New Delhi

For information address:

SAGE Publications, Inc.
2111 West Hillcrest Drive
Newbury Park, California 91320

SAGE Publications Ltd.
28 Banner Street
London EC1Y 8QE
England

SAGE Publications India Pvt. Ltd.
M-32 Market
Greater Kailash I
New Delhi 110 048 India

Printed in the United States of America

Library of Congress Cataloging-in-Publication Data

Main entry under title:

Economic restructuring and political response / Robert A. Beauregard, editor.
p. cm. — (Urban affairs annual reviews ; v. 34)
Bibliography: p.
Includes index.
ISBN 0-8039-3184-0. ISBN 0-8039-3185-9 (pbk.)
1. Urban policy—United States. 2. United States—Industries—Location. 3. Employment stabilization—United States. 4. Industrial organization—United States. I. Beauregard, Robert A. II. Series.
HT108. U7 vol. 34
[HT123]
307.7′6 s—dc19
[338.6′042′0973] 88-15736
CIP

FIRST PRINTING 1989

Contents

Preface

WITHIN ACADEMIC COMMUNITIES, certain concepts frequently seize the collective imagination, generating an outpouring of scholarly work. Deconstruction in literary theory, structuration and the state in sociology, postmodernism in cultural critiques, Fordism in labor studies, and spatiality in geography are recent examples. Even though these concepts rapidly gain great currency, they do not always clarify and, in fact, usually provoke further uncertainty and disagreement. Part of their allure is their dialectic of revelation and opacity. For contemporary urban political economists, economic restructuring is such a concept.

Economic restructuring has two meanings: one as itself, economic restructuring, and the other, more narrowly, as industrial restructuring. The first describes broad changes in the economy, extending beyond the sphere of production into distribution, finance, governmental relations, and the labor process. The second references transformations in the relative importance of industries, what has come to be termed the manufacturing-service shift (Feagin and Smith, 1987; Peet, 1987: 1-5; Scott and Storper, 1986). Thus one component of the concept delineates an arena for investigation; the other contains a largely implicit rationale for the investigation. The reason industrial structure and the economy require scrutiny is that each has undergone profound change. Dynamics and equilibria are not at issue; rather, it is qualitative discontinuities and crises that bother (Bradbury, 1985). The pervasive and temporally continuous relationships that give the capitalist economy its very nature are in flux; not simply a realignment of interindustry linkages, but a shift of the economy away from goods production and into service production, away from assembly line standardization and into flexible accumulation, away from high-wage full-time employment and toward low-wage part-time work. The thematic undercurrent is one of razor-edged cleavages in the processes and structures of capitalist society.

The concept of economic restructuring is relatively new: the late 1970s. After decades of postwar growth in the United States—rising wages, massive job generation, and international economic and political hegemony—an unexpected deep recession occurred between 1973 and 1975. The Organization of Petroleum Exporting Countries struck an economic and symbolic blow to the U.S. economy. Gasoline prices rapidly escalated and with them all the myriad goods made of petroleum. Whereas earlier business cycle fluctuations had involved either high unemployment or a bit of inflation, but not both simultaneously, inflation was rampant and unemployment rates soared. Stagflation was the norm. The recession reverberated through the economy with an emphatic crash as manufacturing plant after manufacturing plant shut its doors or laid off most of its workers. Steel and automobile assembly were particularly hard hit.

Cities felt the recession as fiscal crises. New York City virtually went bankrupt, while other cities balanced their budgets with mirrors. The older industrial cities had been in decline since the early 1950s, some since the early 1930s, but the fiscal crisis was a seemingly new phenomenon. By the latter part of the 1980s, even with a long expansionary period of business activity between 1982 and 1987, the economy was still a hodgepodge of growth and decline. Unemployment rates dropped to a 15-year low and business was booming, but the federal deficit was gigantic, trade imbalanced, the dollar unstable, and the stock market convulsive.

Urban political economists were rocked by these changes; the context for their work was being disrupted. Postwar prosperity had hidden unsettling and basic shifts not only within the U.S. economy but across international economic and political relations. Neoclassical theories of urban and regional development no longer seemed applicable and mere elaborations held out little promise of theoretical breakthroughs much less novel insights into prevailing conditions and processes (Castells, 1977; Harvey, 1973). The scholarly response was to proclaim economic restructuring.

Despite the appearance of economic restructuring in a multitude of scholarly writings, movement to clarity has been arduous and slow. The aim of this book is to push us further along that path, even while uncovering interesting side trails whose contours are more perplexing than comforting. Two themes dominate. The first concerns our theoretical understanding of economic restructuring, both as that understanding corresponds to actual qualitative transformations in the capitalist social formation, and as it reflects upon the usefulness of the concept itself. The second concerns the realm of action rather than that of meaning (Marris, 1987). It poses a different question. How is it that people in their daily lives interpret and respond to economic restructuring? Even if the concept does not enter their vocabulary, does economic restructuring play a role in political action?

Chapters 1 through 3 address the first theme. Susan S. Fainstein and Norman I. Fainstein, in their "Technology, the New International Division of Labor, and Location: Continuities and Disjunctures," review the outlines of the social, economic, and political transformations that have engendered an awareness of economic restructuring. They ask the question of the skeptic: Does scholarly agreement on the existence of a recent dramatic transformation of the U.S. capitalist system confer credence on a profound restructuring, or are we witnessing instead a recent recognition of long-standing trends?

Their approach, like that of other authors in this volume, is strongly spatial. The Fainsteins address economic restructuring in terms of the simultaneity of decentralizing and consolidating trends in specialized economic functions. Their scope, however, is not simply economic, for they range across the societal spectrum looking at pivotal dimensions of the social formation: urbanism, household relations, the social relations of production, and the state. The chapter examines the broad sweep of postwar capitalist restructuring and does find significant transformations. They answer their question in the negative, however: Capitalism has not suddenly restructured. Rather, capital has engaged in a century-long weakening of technical constraints on productive activity and continuously utilized space as "an instrument in the process of creative destruction and worker control." There is no radical discontinuity.

Nonetheless, even if the capitalist economy has continued on a previously defined trajectory and the social formation as a whole has not changed significantly, particular localities have experienced sharp disjunctures. Places like Youngstown, Ohio; Detroit; and Mahwah, New Jersey, faced abrupt plant closures in the late 1970s, and their residents confronted precipitous and sudden declines in household income as massive job loss extended throughout the local economy (Bluestone, Harrison, and Baker, 1981). Within a number of large, central cities, gentrification—the movement of relatively affluent households into less affluent neighborhoods, the subsequent purchase and rehabilitation of housing, and the displacement of longtime residents—also emerged as a by-product of industrial restructuring and exerted its consequences in rapid fashion (Smith and Williams, 1986). In the 1980s, office construction boomed in both central cities and suburbs, and producer services emerged as dominant in many local economies. Thus in a large number of localities, economic restructuring did not present itself as a slow transition or a cosmetic variation on past practices, but as extreme shifts in social and economic relations.

In "Pittsburgh in Transition: Consolidation of Prosperity in an Era of Economic Restructuring," Gordon L. Clark explores the industrial discontinuities within one of the areas hardest hit by plant closures. For those living in Pittsburgh, as Clark points out, the words *transition* and *restructuring* are too antiseptic to describe adequately the changes of the past ten years. Steelworkers, middle-level managers, and even some corporate executives experienced drastic declines in their livelihoods and well-being. Local and national economic stagnation, competition from overseas imports, relocation of production, and the resultant job loss dealt many residents a massive blow. The steel mills along the river valleys have been abandoned, devastating the communities that depended upon them. The city of Pittsburgh also has been transformed from an internationally oriented industrial city to a regionally oriented service economy. In the future, Clark suggests, Pittsburgh will be described by its historical legacy and its residual physical landscape; its buildings and factories monuments to a previous era. Few mills will be left, save perhaps one or two that are preserved as historic sites. The landscape will be more pleasant, less industrial, and more in keeping with other, similar-sized cities around America. Such changes do not fit comfortably within a concept of economic restructuring whose thrust deemphasizes qualitative transformations.

A similar conclusion can be drawn from a study of the consolidation of producer services in global cities, the downsizing of the manufacturing sector, and the expansion of informalization in major cities throughout the United States. In Chapter 3, "New Trends in the Sociospatial Organization of the New York City Economy," Saskia Sassen explores the novel relationships that have emerged to link these disparate transformations. New York City has become a global city, Sassen argues, because of the ability of producer services to consolidate in space while simultaneously extending their lines of control and coordination across political boundaries. Though its process parallels the downsizing of the manufacturing sector, and producer services have indirect effects on manufacturing operations, each is a distinct and separate mode of economic organization. Nonetheless, they conflict in space through attempts to impose their own logic on the use and value of the built environment. Manufacturing zones become transformed into residential loft districts and office complexes encroach upon working-class neighborhoods. Immigrant communities, their inhabitants attracted in part by the "global" image, support the producer services sector and its highly paid labor force with an array of informal economic activities that provide distinctive and specialized services and commodities for the affluent. Skilled immigrants reconstitute a new, less visible, and smaller-scale manufacturing sector, and unskilled immigrants join the ranks of low-paid service workers. The New York City of global pretensions is thereby restructured spatially and socially as broad economic trends vie for dominance.

The immigrant inhabitants of global cities often find themselves exploited within sweatshops and entangled in the legal web of punitive immigration laws. The downsizing of the manufacturing sector creates large numbers of skilled workers whose talents are no longer in demand, and whose incomes no longer support the life to which they have become accustomed. The retail, consumer, and distributive services that cater to the financial, legal, and administrative firms generate numerous jobs but at low wages, further contributing to the social and economic polarization of the city. These consequences of economic restructuring, however, have yet to manifest widespread political mobilization. Immigrants channel their energies toward the interstices of the formal economy, displaced workers slide into retirement or service sector jobs, and ill-paid service workers move in and out of the labor force as high turnover rates create family instability and undermine labor organizing.

The lack of widespread social unrest poses a theoretical conundrum. Certainly capital has not flinched from shifting the cost of economic restructuring onto workers. The U.S. automobile industry, for example, weathered foreign competition and managerial excesses and incompetence through massive labor cutbacks, tough contract negotiations resulting in give-backs and wage reductions, joint ventures with foreign firms, and "offshore platforms" for assembly (Hill, 1988; Clark, 1986). Despite these and numerous other onslaughts, labor has remained relatively passive. Certainly deep trenches separate the workplace from civil society (Katznelson, 1983), but the divide is bridged by the workers who occupy both. As Piven and Cloward (1977: 21) have written, "it is the daily experience of people that shapes their grievances, establishes the measure of their demands, and points out the targets of their anger." What more salient daily experience exists under capitalism than the experience of the workplace? Protest, nonetheless, requires a transformation of consciousness and the production of collective defiance. Economic restructuring alone thus seems insufficient as a cause for mass political action (Fainstein and Fainstein, 1985).

Chapters 4, 5, and 6 explore the complex linkages between economic restructuring and political activity. Each attempts to connect collective actions undertaken to reconstitute power and influence within a spatial realm with shifting economic conditions. Ann Roell Markusen, in "Industrial Restructuring and Regional Politics," argues that industrial composition is one key to a region's political temperament. As regions confront one or another type of upheaval emanating from rapid and profound restructuring, that temperament is changed. Regions with an aging industrial complex, where the profit squeeze has produced wholesale plant shutdowns, face one set of problems, while those with rapid growth from high-tech defense industries or relocating mature industries confront another. More specifically, older mature regions with a set of declining industries—Pittsburgh's region, for example—are apt to generate internal solidarity around a common set of regional goals and to levy demands upon a higher level of government. The national government is frequently blamed for failing to control interest rates, allowing international exchange rates to erode demand for the region's export commodities, and avoiding appropriate restraints of foreign subsidization of import competition. Regions like the Old South, on the other hand, are the recipients of profit-squeezed industrial capital attracted by probusiness regulatory climates. Here, internal conflict develops between incoming and established business elites, between elites and workers with growing class consciousness, and between environmentalist and progrowth advocates.

Regions hosting high-tech growth, for example, Silicon Valley, experience numerous environmental and governmental problems as population and investment quickly expand. Political conflict breaks out between outright boosters and residentialist forces seeking to control the boom and improve the quality of life. The conflicts between production and consumption are pronounced, while intraclass rivalries predominate. Overall, regional politics varies greatly by underlying economic experience, Markusen argues, yet the relationship between economic restructuring and political activity is not unmediated by culture, tradition, and class structure.

In Chapter 5, Robert A. Beauregard uses this overall theme and the more specific relations between production and reproduction, workplace and civil society, and local government and the labor process to address the forces shaping political mobilization in large, former industrial cities. The goal of "Local Politics and the Employment Relation: Construction Jobs in Philadelphia" is to explain how a specific instance of grass-roots political activity—one focused on the employment relation—was created and mediated by economic restructuring, evolving modes of state activity, and the sociospatial transformations of postwar Philadelphia. The event is the attempt by a variety of community-based organizations in the mid-1980s to have the city council pass an ordinance that would give hiring preference to city residents, minorities, and women on governmentally subsidized construction projects. Because the equal employment laws of the federal government and the broadly recognized discriminatory practices of the construction industry are central to the emergence of this political event, the argument focuses on racial divisions.

Beauregard concludes that the economic restructuring that is reshaping the industrial composition and labor markets of Philadelphia affected this specific instance of political mobilization primarily in an indirect fashion: specifically through its effect on migration and the accumulation function of local government. The civil rights stance of the federal government, the emergence of a black electoral coalition, and the increasingly active prodevelopment orientation of the local government have been of much greater moment in structuring the activities of the ordinance's supporters and the government's responses. Such findings, based as they are upon a situation merging production and reproduction and thereby drawing the state into the labor process, pose a challenge to prevailing theories of urban social movements that erect conceptual barriers between the workplace and civil society, much as capitalism has separated them spatially and in practice. Thus Beauregard sees economic restructuring creating broad channels for political activity, leaving indeterminate the specific interplay of contending forces as political demands are tendered, deflected, reconstituted, and absorbed.

In "Spatial Change and Social Justice: Alternative Economic Development in Chicago," Robert P. Giloth and Robert Mier shift attention even more to the place-specific nature of community development and political organization. They argue that the recent upheaval in inner-city industrial and working-class areas caused by national and international economic restructuring is not beyond the control of local planners nor immune to the actions of community-based organizations. To believe otherwise is to isolate political movements from local communities, and to forgo opportunities for effective intervention.

Using as illustration the changing organization of production in Chicago—specifically in such sectors as steel, apparel, printing, and transportation—Giloth and Mier point out the complexity and contingency of industrial restructuring. The end result of the simultaneous concentration and deconcentration of production, the bifurcation of the labor market, and the polarization of income is not an orderly arrangement of investment, but chaos and irrationality in the use of urban space. The goal of a progressive politics, then, is to shift attention from indeterminant and unpredictable spatial and economic artifacts to more enduring human relations. A politics of human relations, however—one oriented to social justice as fairness and equality of opportunity—must in turn be rooted in local politics. For Giloth and Mier, then, economic restructuring is having a profound impact on Chicago, but in the interstices of its discontinuities exist numerous opportunities for political action to achieve social justice.

What are we to make of these various arguments about economic restructuring and its relation to politics? Robert A. Beauregard, in "Space, Time, and Economic Restructuring," our final chapter, places these postwar transformations and their theoretical elaboration in a historical and spatial context. He argues that economic restructuring is inherently elusive, a function of its uneven temporal and spatial development. Time and space refract economic restructuring into a multitude of disparate phenomena whose common origins are lost in the particularities induced by uneven development. While economic restructuring projects relatively coherent characteristics if broadly formulated and is clearly exhibited in global cities, its processes and consequences are diffused as we focus on specific localities and instances. The notions of structure and periodization are key to understanding these complex phenomena, and Beauregard uses them to elaborate on his general theme.

Subsequently, he applies this variegated perspective to the nature of the political response to economic restructuring in the United States. While restructuring imposes burdens on labor and creates conditions that would seem to engender labor's political resistance, restructuring also unfolds in such a way as to undermine collective consciousness, mass mobilization, and progressive political action. Fragmented and reactionary responses remain. They do little damage to capital's hegemony and provide minimal relief from the costs borne by labor.

In this manner, Beauregard returns to the issue of conceptualization. The various theoretical dimensions of economic restructuring are explored in order to understand better its causes and consequences. Without doubt, economic restructuring is a useful theoretical probe for analyzing and comprehending the workings of contemporary capitalism.

The flexibility of its theoretical meanings, however, serves as both a source of strength and a source of weakness. The concept must not be allowed to devolve into a faddish symbol of intellectual relevance. The initial rush of enthusiasm, undoubtedly, has produced many worthwhile studies, and revealed relations not heretofore considered important. The challenge for theorists, as always, is to combine a critical self-reflection with an expansive application. Only then will economic restructuring continue to yield insights, and only then will comprehension, as it meanders through the discourses of society, engender the appropriate political response.

—Robert A. Beauregard

REFERENCES

BLUESTONE, B., B. HARRISON, and L. BAKER (1981) Corporate Flight. Washington, DC: Progressive Alliance.

BRADBURY, J. H. (1985) "Regional and industrial restructuring processes in the new international division of labor." Progress in Human Geography 9 (1): 38-63.

CATELLS, M. (1977) The Urban Question. Cambridge: MIT Press.

CLARK, G. L. (1986) "The crisis of the midwest automobile industry," pp. 127-148 in A. J. Scott and M. Storper (eds.) Production, Work, Territory. Boston: Allen & Unwin.

FAINSTEIN, S. S. and N. I. FAINSTEIN (1985) "Economic restructuring and the rise of urban social movements." Urban Affairs Quarterly 21 (1): 187-206.

FEAGIN, J. R. and M. P. SMITH (1987) "Cities and the new international division of labor: an overview," in M. P. Smith and Joe R. Feagin (eds.) The Capitalist City. Oxford: Basil Blackwell.

HARVEY, D. (1973) Social Justice and the City. Baltimore: Johns Hopkins University Press.

HILL, R. C. (1988) "Global factory and company town: the changing division of labour in the international automobile industry," in J. Henderson and M. Castells (eds.) Global Restructuring and Territorial Development. London: Sage.

KATZNELSON, I. (1983) City Trenches. Chicago: University of Chicago Press.

MARRIS, P. (1987) Meaning and Action. London: Routledge & Kegan Paul.

PEET, R. [ed.] (1987) International Capitalism and Industrial Restructuring. Boston: Allen & Unwin.

PIVEN, F. F. and R. A. CLOWARD (1977) Poor People's Movements. New York: Pantheon.

SCOTT, A. J. and M. STORPER [eds.] (1986) Production, Work, Territory. Boston: Allen & Unwin.

SMITH, N. and P. WILLIAMS [eds.] (1986) Gentrification of the City. Boston: Allen & Unwin.

Acknowledgments

THIS BOOK OWES its origins to a lecture series held during the 1985 to 1986 academic year at the School of Urban and Regional Policy, Rutgers University. The funding came from a variety of university sources: the Graduate School, the Faculty of Professional Studies, and the Provost's Office. The person responsible for arranging the financing for the series was Dr. Salah El-Shakhs, then director of the School of Urban and Regional Policy (SURP). I owe him a special debt. Sunny Hamburg, SURP's administrator, helped me to arrange and publicize the lectures and to care for the visiting lecturers; and the Department of Urban Planning and Policy Development provided a variety of support services. Ann West, the editor at Sage Publications, offered early support for the project and shepherded the book from prospectus to distribution. To all of these people, and to others whose contributions were less specific, though essential, I give my heartfelt thanks.

1

Technology, the New International Division of Labor, and Location: Continuities and Disjunctures

SUSAN S. FAINSTEIN
NORMAN I. FAINSTEIN

RECENT SCHOLARSHIP has converged to produce surprising agreement on the characteristics of the contemporary capitalist spatial system. The common perception is of highly decentralizing tendencies among both production and consumption units along with the clustering of certain specialized functions in nodes. Spatial decentralization is paradoxically made possible by economic concentration: "We find a tendency both toward the delocalization of the logic of the economic process and toward the concentration of decision-making units in a few commanding heights of the international economy" (Castells, 1985: 30). The great size of the modern corporation permits it to break up its component parts into separate operating units. The large costs of financing giant enterprises, the negotiations involved in acquiring and recombining them, and the necessity of coordinating their far-flung parts require spatial concentration for deal-making and control: "Key cities throughout the world are used by global capital as 'basing points' in the spatial organization and articulation of production and markets" (Friedmann, 1986: 71; see also Sassen-Koob, 1986).

The causal mechanism underlying spatial patterns is likewise a matter of considerable consensus despite varying emphases. Basically it involves two interactive conditions: (1) the freeing of industry from

---

AUTHORS' NOTE: *We wish to thank Hooshang Amirahmadi and Robert A. Beauregard for their helpful comments on an earlier draft of this chapter.*

locational constraints as a consequence of technological innovation, and (2) the organizational imperative of capitalist industry to achieve conditions of profitability through reducing factor costs. Thus an increasingly footloose capital can take advantage of lower labor costs, market proximity, and favorable governmental policy (a "good business climate") with relatively little regard to the forces that brought it to a location in the first place:

> With its wide array of locational choices, but with often very little dependence on any particular locality, the multi-establishment firm is able to make widely separated hinterlands compete with each other for development; in recent years an interregional and international bidding war for plant locations has been effectively ignited [Scott and Storper, 1986: 306].

Castells (1984, 1985) particularly stresses the role of technology both as a facilitator of mobility for existing economic activities and, through the rise of new high-tech industries, as a direct engine of growth and decline in particular regions and metropolitan areas.

This depiction of spatial behavior differs from traditional location theory not so much in its deemphasis of agglomeration economies and transportation costs, both of which remain significant variables in a firm's ultimate locational decision, as in its systemic viewpoint (see Markusen et al., 1986; Massey, 1984). Analysis has shifted from the locational decisions of particular firms and industries, or the attractiveness of a specific metropolitan area, to the movement of capital and labor throughout the entire world system of cities and regions (see Harvey, 1973: chap. 6). The strategies of capitalists as well as the activities of the state in affecting those strategies become the most important variables underlying spatial change. The outcome to be explained is not the growth or decline of a place but uneven development across places.

The new approach to understanding spatial outcomes derives from the introduction of Marxist political economy into the study of historical geography. This chapter raises the question of whether novelty lies simply in how the problem is perceived, or whether the processes being described are also new. In particular, we wish to examine whether the post-1973 restructuring of the capitalist system did usher in a new period of spatial arrangements or whether the present epoch is continuous with the past.

Many of the leading theorists who describe capitalist spatial structure see a sharp disjuncture resulting from the capitalist crisis of the 1970s. Thus Castells contends:

> The most important global process conditioning the relationship between new technologies and spatial dynamics is the economic restructuring that U.S. capitalism is currently undergoing, superseding the structural crisis of the 1970s [1985: 12].
>
> It is mainly through the new economy that high technology is deeply modifying our cities and regions. . . . This perspective assumes that the world economic system, still centered in the United States, is in a process of deep structural transformation, in connection with the technological revolution (although *not* determined by it), and that only by understanding the effects of technology within this broader framework (we call it the economic restructuring of the world capitalist system) will we be able to assess the impact of high technology in the new urban-regional process [1985: 19].

Harvey (1987a: 7) pinpoints the decisive moment of demarcation between the previous era of "Fordism" (Liepetz, 1986)—characterized by postwar boom and high mass consumption levels within the advanced capitalist countries—and the present situation:

> The recession of 1973 shattered that [earlier] framework and initiated a process of rapid and as yet not well understood transition to what appears to be *an entirely different regime of accumulation* [emphasis added].

Similarly Storper and Scott (1986: 8) assert:

> The new historical realities of capitalism began with a crisis of profitability in production in the developed nations. This same crisis created a deepening internationalization of production and financial systems. Through the interaction of industrial production and money capital, internationalization has now emerged as a relatively external pressure on the organization of production, work, and territory in all countries.

Jameson (1984: 80), in an extraordinarily influential essay on the culture of late capitalism, refers to "the whole new decentered global network of the third stage of capitalism itself." He traces the sprawling, unfocused style of the modern city and the recently constructed megabuildings within it, epitomized by architect John Portman's atrium hotels, to a new multinational stage of capitalism. According to his analysis, the "postmodern" period is "the purest form of capital yet to have emerged, a prodigious expansion of capital into hitherto uncommodified areas" (Jameson, 1984: 78).

Despite the widespread consensus that the last 15 years have witnessed a mutation in spatial forms attributable to changes in the underlying relations of production, arguments about how discontinuous the process really has been contradict each other. Thus Scott and Storper (1986: 306) refer to peripheral decentralization occurring over "the last several decades" and the breakup of firms into geographically separated segments "over the course of the twentieth century." Jameson is fuzzy as to whether postmodernism dates back to the end of World War II (the beginning of Mandel's [1975] period of late capitalism and the start of Harvey's period of Fordism) or commences in the 1970s with the demise of modernist culture.

Harvey's assertions concerning historical discontinuity depend on the accuracy of his characterization of the 1945-1973 period as dominated by capital-labor relations in geographically stable mass production industries; under this analysis, the competitive and service sectors recede in importance during that era. Yet he (1987a: 6) admits that "Fordist wage bargaining was confined to certain sectors of the economy where stable demand growth could be matched by large-scale investment in mass-production technology. Other sectors of high-risk production still depended on low wages and weak job security." O'Connor (1973), in fact, has argued that the competitive sector, rather than being a declining remnant of early capitalism, functioned significantly throughout the postwar era to allow adjustments to fluctuation in consumer demand and the need for labor. If he is correct, then accumulation before 1973 was far more flexible than Harvey's disjuncture argument implies.

For Harvey (1986: 51 ff.) and others, the concept of a break in the evolution of capitalism leads to an identification of the present time as involving a new integration of the processes of production and consumption in which capitalists seek to resolve crisis through a "spatial fix." It is our contention that the processes described have largely characterized capitalism for most of this century, and that, while discontinuities exist, they do not justify calling the post-1973 period a new stage of development. Our method of exploring this issue is first to describe the tendencies that have seemingly accelerated the process of spatial transformation and to establish their continuities with the past.

We then analyze their impacts to determine the significance of changes in the spatial system.

TENDENCIES IN WORLD CAPITALISM

The drive for profitability, as discussed above, leads capitalists constantly to seek lower factor costs. N. Smith (1984: 157) argues that, since 1973, we have been in a period of devaluation of fixed capital and labor power. This devaluation expresses itself geographically as territorially differentiated investment and disinvestment. But Smith calls this the typical capitalist response to crisis. Newness is not then in the nature of the response but in the types of locales in which capital is invested and disinvested. At the urban scale, investment moves into the suburban periphery or previously devalorized core areas while withdrawing from old manufacturing centers. Worldwide, capitalists establish new manufacturing centers in cheap labor areas. Two of these phenomena represent a seeming departure from the past: revalorization of the urban core; and the establishment of foreign branch plants or subsidiaries that export their products either back to the country of ownership or to other places rather than serving a local market (Feagin and Smith, 1987: 10).

These are not, however, unprecedented breaks. As Harvey (1985) and others have shown, Haussmann's Parisian effort in the 1850s set the model for urban revitalization. The movement of production to cheap labor areas and the closing or absentee ownership of factories as a consequence of mergers have been features of manufacturing since the 1880s. McKelvey (1963: 42) notes that "so many [American] cities suffered [before the passage of the Sherman Anti-Trust Act in 1890] from the migration of old companies and other effects of consolidation that the antitrust forces won general approval." Waldinger documents the flight of the garment industry from the northeastern United States in the 1920s (Waldinger, 1986: 54-56); and Perry (1987: 119, 134) describes Buffalo, New York's, transformation into a dependent economy at the turn of the century as absentee owners—either through the construction of branch plants or the acquisition of previously independent establishments—took over its manufacturing enterprises. McKelvey (1963: 43) chronicles the post-1900 suburban migration of factories that stimulated the rise of urban chambers of commerce devoted to attracting and retaining business in a fierce interlocality competition for economic development.

Indeed, decentralizing tendencies have existed throughout the history of capitalism, even at the apogee of the nineteenth-century European manufacturing city.[1] Internationalization of capital started with the great trading companies of the seventeenth century; the movement of industry to urban peripheries began with the development of the railroad. The company town and the branch plant represent efforts by capital throughout the twentieth century to reduce factor costs through availing itself of cheaper land, more malleable governments, and less militant labor forces (Gordon, 1978: 47-51). The telephone, the truck, the automobile, and the airplane made possible the dispersal of home and industry and the separation of headquarters from production facilities that has characterized much of the twentieth century. Surely the computer and telecommunications are only the latest step in an ongoing process whereby physically disaggregated units are tied together through transport and communications mechanisms.

Moreover, the deindustrialization of the United States and Western Europe is traceable less to the greater capability of domestic capital to produce offshore than to the rise of new industrial powers in the Far East. Here we see a pattern that is new in terms of the particular states acquiring dominance but old in the history of capitalism, which has witnessed a number of shifts in the locus of production. Beginning with the decline of the Italian and Dutch cities relative to Britain, and continuing through the rise of the United States and Germany as manufacturing powers, market displacement has periodically followed years of stable hegemony. The assimilation of previously nonindustrialized countries like Sweden and Japan into the capitalist core states has occurred frequently throughout the history of capitalist development (Wallerstein, 1980).

Tied to arguments concerning spatial deconcentration are hypotheses about the new international division of labor (NIDL). As stated by Frobel et al. (1980), the NIDL refers to the decomposition of production processes to allow capital to exploit production sites throughout the world. The qualitative change here is not in the use of poor countries as producers of cheap goods—their status as providers of unfinished raw materials always placed them in this position—but in their role as manufacturers of finished goods. In the core countries, as discussed above, the NIDL causes the abandonment or transformation of manufacturing areas. In the Third World, the introduction of modern manufacturing techniques produces overurbanization, as agricultural populations are attracted to cities whose capital-intensive industry cannot absorb their labor power. According to Sassen-Koob (1987:

chap. 4), the NIDL also stimulates the emigration to Western Europe and the United States of Third World workers who have become familiar with Western culture through the presence of Western industries and aid missions; these new immigrants settle in highly concentrated patterns, especially in the burgeoning global cities.

Certain aspects of these processes are not strongly deviant from the past. The migration of agricultural workers to urban centers as a consequence of the capitalization of agriculture began with the British enclosures and continued through the rise of German and American industry. While the absorption powers of the early industrial cities were greater than those of contemporary Third World metropolises, those cities nevertheless saw frequent periods of high unemployment and misery. The tapping of lower-cost, free labor pools was pioneered by England and the United States in the nineteenth century, primarily through importing surplus labor rather than using it in place. Emigration from poor to well-off nations thus has been a continuous process throughout capitalist history, although the sources of labor have shifted and more countries have become involved (Sassen-Koob, 1987: chap. 4). The growth of manufacturing sectors in poor countries as a consequence of direct foreign investment does represent a growing recent phenomenon with important consequences for both investing and host nations. But, as both Sayer (1986) and N. Smith (1984) assert, continuation of this trend is not an inevitable consequence of capitalist logic and may well taper off for both political and economic reasons.

Furthermore, manufacturing location is only one component of the spatial structure of production. The geography of the advanced countries for much of this century has been strongly determined by their service sectors (see Mingione, 1981: chap. 1). Within the United States, position in the hierarchy of cities has long depended on success in attracting service industries (Noyelle and Stanback, 1983). Service employment (including wholesale and retail trade, finance, insurance, and real estate [FIRE], government, and miscellaneous services) exceeded goods-producing and transportation employment by the year 1919 (U.S. Bureau of the Census, 1960: table 73). Manufacturing employment increased by only 43% between 1945 and its 1980 peak, while service employment skyrocketed by nearly 200% in the same period from a larger base (U.S. Bureau of the Census, 1985a: table 683). Thus the overwhelming importance of services in determining the economic well-being of metropolitan areas is an old phenomenon; albeit the change from slow growth to no growth in the manufacturing sector may represent a significant transformation.

In summary, then, throughout the history of capitalism and its periodic restructurings, we can identify a number of forces of "creative destruction" that manifest themselves spatially as well as within social relations and that in turn affect social and productive relations. These include the devalorization of fixed capital, geographical disaggregation of the integrated firm, the flight of industry to areas with lower factor costs, absentee ownership, the entry of new states into the capitalist core, new international divisions of labor, and the increased importance of service-and-control employment within the occupational structure (see Smith and Tardanico, 1987). While some of these forces have recently become more prominent, almost all have predated the crisis of capital in the 1970s.[2] The question of whether we have entered a radically different stage then depends not on the existence of new forces but on whether the impacts of these mainly continuing forces, heightened by recent events, are such as to create a new spatial configuration.

SPATIAL FORMS IN WESTERN EUROPE AND NORTH AMERICA

A qualitatively distinctive spatial organization emerges when underlying forces are sufficiently strong not only to create change along a series of relevant spatial dimensions, but also to initiate new forms of consciousness, social practices, conflicts, and solidarities, as spatial change and social change dialectically reinforce one another. From this perspective a new type of spatial configuration necessarily must be accompanied by new forms of social relations (see Harvey, 1987b: 13). In other words, the more extensive the social transformations interacting with spatial reorganization, the more likely we are to be witnessing a historical disjuncture. It is such a disjuncture that in turn makes a series of perhaps gradual, quantitative spatial rearrangements add up to a new type.

In order to determine whether recent developments constitute an emergent type of spatial organization, we examine spatial and social rearrangements simultaneously along some of the major dimensions that define a social formation: urbanism, the household, the social relations of production, and the state. These characteristics, however, vary substantially within the core states and even more within the world economy of capitalism as a whole. While it is possible to identify general forces at work worldwide, to the extent their spatial impacts are significantly mediated by the physical and social terrain of particular places, we must necessarily limit our analysis to a relatively homogeneous set of places. In the discussion that follows, we concentrate on Western Europe and North America, recognizing that in the process we largely avoid the question of whether a qualitatively new spatial organization has emerged in the capitalist semiperiphery and periphery. Enough claims, however, have been made about transformations in the core states of Europe and North America to make their examination theoretically interesting, even if the generalizations we can make must be confined to these places.

URBANISM

The character of *urbanism*, by which we mean the form and function of urban agglomerations, is initially established by the organization of production, both locally and on a larger scale. We have already enumerated the forces at work on a worldwide basis that are producing changes in the spatial organization of production in many different places. *In general, we may say that these forces are reducing the technical necessity for a particular form of urban development, rather than imposing an emergent type.* Most strikingly, the manufacturing city has become obsolete in the advanced capitalist nations, just as did the "military" city with its wall and defenses a few centuries earlier. Nineteenth-century patterns, defined initially by the technical necessity of factory workers being able to walk to work and later to take mass transportation, remain reified in the built environment. But they are no longer reproduced by the spatial necessities of factory production or of transportation technology. Even on a regional scale, the relation between production sites and workers' housing is relaxed, and a variety of new forms have emerged. In some parts of Italy, for instance, factory workers live in rural villages, commuting long distances to urban manufacturing districts (a pattern that Konrad and Szelenyi, 1977, call underurbanization). In Britain, modern manufacturing production (along the M4 vector) has been organized spatially within a planning practice that sharply demarcates town and country. The situation in the United States, in contrast, is most usually one of suburban sprawl, with neither plants nor housing organized into coherent development patterns.

The great heterogeneity of service-producing industries means that various sectors must be considered separately to allow any sort of generalization concerning spatial dynamics. Producer's services, particularly investment banking and corporate law, are the least decentralized industry of the advanced countries and the major contributor to the global city phenomenon; as David Vogel (1986) indicates about New York:

> Regardless of the geographic base of the current generation of corporate raiders and speculators—which in fact appears to be rather diverse—or of the consequences of their efforts on the physical location of corporate headquarters within the United States—which is so far unclear—there can be no question that the current wave of mergers has been an enormous boon to New York City's financial service sector: mergers and acquisitions are extremely profitable. Physical proximity does seem to matter in these negotiations and anyone interested in either defending or instigating a corporate takeover has no choice but to enlist the services of investment bankers and corporate lawyers in New York City.

Thus despite new technologies that permit negotiations at a distance, organizational imperatives continue to reinforce concentration.

Office work more generally, however, is following a route quite similar to that traversed earlier by manufacturing (see Nelson, 1986). Whereas the spatial deconcentration of manufacturing was facilitated by the development of telephone and transportation technology, allowing communication between headquarters and production sites and the movement of goods to far-flung markets, the decentralization of office work awaited breakthroughs in telecommunications technology for data transmission. The effect has been a further reduction of the necessity of central cities for economic functioning and therefore an important reinforcement of the ongoing process of decentralization in the advanced countries. Even in the relatively centralized financial services industry in the United States, decentralization of back-office operations is occurring:

> There has been a gradual decoupling of front and back office operations in most information intensive firms. The back offices, with their specialized requirements, have been moved out of prime central city real estate to lower cost locations, either to the periphery of metropolitan areas or to regions that offer comparative advantages in labor, energy, taxes, and/or amenities. The locational choice is frequently a product of corporate strategy; some firms want to avoid white collar unions; other firms want computer operations to be in close proximity to headquarters in order to have easy access to technical staff [Moss, 1986: 40].

Consumer services like restaurants, laundries, or branch banks must be located proximate to their markets. Hence residential decentralization, a long-term trend, rather than the drive to bring down operating costs, determines dispersion of consumer services. The separation of home from work, which commenced with factory production, stimulated the initial stages of this type of decentralization (see Katznelson, 1981). Post-World War II suburban expansion accelerated the dispersal of retailing. The decline of downtown retail centers began with the growth of suburban shopping centers. In those advanced countries where planning regulations constrain suburban sprawl, core areas retain their importance for retail trade.

Other consumer services like insurance carriers or credit card operations can be located virtually anywhere; their decentralization preceded the economic restructuring of the 1970s-1980s and continues gradually. Mail-order retailing, carried on from obscure locations, appears to be a recent consequence of telecommunications and computerized inventory control, but its antecedents are the Montgomery Ward and Sears Roebuck catalogues, which served vast rural hinterlands and were based in locations far from their customers. What has changed as a consequence of the changing household is the nature of its market. Rather than serving isolated farm families, retailers cater to two-earner households where the inhibition on shopping in stores is lack of time rather than geographical access.

Castells (1985) describes a set of apparently contradictory trends made possible by telecommunication and data-processing technology. Production may move again into the home in new electronic cottage industries, as workers sit glued to their CRTs, their output measured in a central headquarters computer. At the same time, some unique places characterized by attractive architecture or geography become almost priceless locations for high-income residents, although there is little evidence of a shift in preferences among the Western upper classes as a whole toward dense urban living. Most places become increasingly homogenized, simply defined as nodes through which communications flow. Castells goes so far as to suggest a "real revolution," with families and individuals plugging into satellites and computers for their entertainment. In this view, we may even be witnessing a disintegration of urban culture, as "the new technologies lead toward the delocalization of experience in the sphere of private life" (Castells, 1985: 18).

With such a complex picture, it is difficult to pin down Castells's claims. On the whole, however, his image of an electronically transformed or perhaps destroyed urbanism seems exaggerated. For one thing, these tendencies have long been around. The automobile and the television were probably an order of magnitude greater in transforming power than the computer. Three decades ago, sociologists in the United States were claiming (and counterclaiming) that suburbanization was transforming the fabric of American culture. If place-based community has been declining, the process has been a rather steady one, with no recent shift readily apparent. Indeed, powerful forces associated with the NIDL are reproducing the most traditional and place-bound mode of urban occupancy: the immigrant community (Sassen-Koob, 1986; *New York Affairs*, 1987). Foreigners are compelled and choose to live together, whether they be immigrants in the United States and Britain, or guest-workers on the European continent. Vast districts of New York, Miami, Chicago, and Los Angeles increasingly look more like the cities of 1900 than the moonscape predicted or dreaded by Castells.

Moreover, agglomeration economies—or at least the preferences of economic elites for agglomeration—still prevail in many service industries, such as advertising, publishing, broadcasting, law, accounting, and banking. In places where such industries predominate, office production is likely to maintain a big-city form (this is the case, by definition, for the world cities). Even in manufacturing, firms may prefer to concentrate production; Hill (1985), for example, shows how Toyota chose to organize automobile production spatially by creating a company city (although he points out that Nissan's approach was to decentralize). High-technology production, advanced electronic communications, and the NIDL are not in themselves establishing a new urban form. The spatial outcome depends on sociopolitical factors as well as on the differing logics of particular firms rather than on the logic of capital in general. For all these reasons, the evolving spatial patterns of capitalist production and consumption have not, therefore, resulted in a new type of urbanism, although, on the whole, the decentralizing tendencies that began early in the century have continued.

HOUSEHOLDS

The reorganization of the household appears to mark the greatest shift from earlier periods of capitalism, when the home became the isolated haven in which women sustained the men who had to strive within the competitive work world (Zaretsky, 1976). The separation of paid work and home created by urban, industrial life in the nineteenth century sharply curtailed the production functions of the family.

Increasingly, reproductive functions have also departed, as cultural and economic forces interacted to commodify many services previously provided by women (e.g., cooking, child care) and thereby to reduce the time spent by women on (unpaid) domestic work (see L. Vogel, 1983).[3] Far from a reunification of work and household (with all its urban implications), we are witnessing the continued dismantling of pieces of the home, which now come to be purchased by working husbands and wives, more of whom earned their wages (legally or not) in the production of those same services. In most places, there has been an expansion in formally organized, as well as commodified, service production: proprietary businesses sell child-care services, rather than the neighbor down the block. Several factors underlie the process: recent efforts by capital to lower labor costs through female participation, reductions in pay to male wage-earners forcing other family members to work, increases in the number of female-headed households, feminist consciousness, and rising female educational levels.

This line of argument is at odds with emphases on an expanding sphere of household production as a major tendency in capitalist economies. Writing early in the decade, Szelenyi (1981: 9) drew a picture not dissimilar to Castells's imagery of the wired cottage:

> The "domestic mode of production," the informal economy, may no longer be shrinking, on the contrary it may aggressively expand, decommodifying economic activities, taking them out of the "capitalist sphere." This change could be reinforced by environmental problems and the energy crisis. Solar energy systems might lead to decentralized forms of energy production, in which smaller production units, different forms of "domestic units" (which might be communal in organization and not necessarily "nuclear families") might play a role. This would have unprecedented effect on urban and regional systems.

Other analysts took similar positions (see Gershuny, 1978; Pahl, 1980), though often with quite different conceptions of what they meant by household production and informalization. The last decade did not, however, fulfill these predictions (note Pahl's revised interpretation, 1984). A great deal of the "informal" work discovered in Britain seemed to have been a temporary product of unusually high unemployment rather than an indication of structural change. More important, the increase in female labor force participation, undoubtedly associated with more intensive capitalist exploitation of labor, resulted in increased consumer demand for service provision outside of the household.

Changes in the household have affected what happens in particular places, but have not really altered spatial development. Builders scaled down the average size of new housing units, and in many places larger families (typically with lower incomes) are forced to live far from town centers. Modifications in residential structures and commodified consumption have primarily occurred, however, within previously defined spatial patterns. The high-rise apartment building with rooftop health club has its analogue in the horizontal complex of garden apartments with pool and gymnasium in the center. Singles' bars and athletic clubs, therapy groups, and gourmet dining have substituted for earlier familial and communal forms of recreation. The important point, however, is that these functions can be carried out within a great variety of spatial configurations—they do not determine a particular spatial outcome.

THE SOCIAL RELATIONS OF PRODUCTION

Within the social relations of production, the most obvious phenomenon has been the stagnation of employment in goods-producing industries within the old core of capitalist nations. Working-class employment growth in these nations since the 1960s has been confined mainly to office work and services. A substantial portion of workers in Europe and North America has been pushed into the secondary labor market (Piore, 1979) or into irregular labor (Mingione, 1981: 128 ff., 1983). As noted previously, the participation of women in the wage economy has increased to varying degrees throughout the West, especially in the United States. The restructuring of work and of labor force participation has been associated with an overall deterioration of the class situation of both old and new working classes, as indicated by an increase in the level of economic inequality, a decline in the aggregate share of income received by labor versus capital, and an intensification of cleavages based on sectoral position and social grouping within the lower classes.

Sassen-Koob (1986: 91-92) identifies four factors in the job supply that have increased inequality in the distribution of wages since about 1970:

(1) historically lower wages in services than in manufacturing;
(2) relatively high levels of wage inequality within the fastest-growing industries in both goods-producing and services-producing sectors;
(3) a "downgrading" of manufacturing jobs that results from the preponderance of low-wage work in new manufacturing industries like electronics

combined with the reorganization of the relations of production in old manufacturing industries, notably the repression of labor unions along with the rise of sweatshops and homework; and

(4) technical reorganization of manufacturing work involving the disappearance of semiskilled and skilled middle-wage jobs, as computers and robotics replace workers on the shop floor.

Aggregate data on income distribution demonstrate the growth in economic inequality. Thus the GINI coefficient for family income in the United States increased by about 10% between 1968 and 1983 (U.S. Bureau of the Census, 1985b: table 17). New job creation in the United States after 1970 was mainly at the low and high end of the wage distribution, thereby producing an increasingly bifurcated pattern (Bluestone and Harrison, 1986; Lawrence, 1984; table 1; Sassen-Koob, 1986: table 1). During the 1980s, profits increased much more rapidly than wages as capitalists successfully revalorized the returns to ownership and drove down the returns to labor. Moreover, in the United States at least, social group inequalities have been maintained or increased, thereby adding widening gaps within class categories. Despite the women's movement and the entry of many women into managerial jobs, women's wages have remained substantially lower than male wages. The overall economic situation of blacks has deteriorated sharply, including the position of the black middle class in relation to its white counterpart (N. Fainstein, 1987). Immigration has further contributed to divisions within the working class. The sharp economic decline of the Great Lakes manufacturing region, the so-called Rustbelt, has added an additional, territorial, dimension to inequality.

While national histories and social structures mediate economic restructuring, the typical effect in the core states of Europe and North America has been not only to increase the cleavage between capital and labor, but also the fragmentation of labor:

> The working class is more and more disunited because the working and living conditions of its various strata are more and more differentiated: from the trade-unionized, highly paid blue-collar worker, employed permanently in a large corporation, or the white-collar worker, regularly employed by the state, the large banks and insurance companies, down to the unskilled day-laborer irregularly employed in small building industries, the female home-workers, semi-employed by small textile units, the retired person, working part-time and irregularly to supplement a very low pension, and so on [Mingione, 1981: 129].

Nevertheless, fragmentation of the lower classes has been a persistent characteristic of capitalist social structure and is now rather less extreme than was the case at the beginning of the century. At that time, the working class was severely split between agricultural and manufacturing workers, and in the United States immigration-produced ethnic differentiation was even more divisive than now. Seen from the perspective of the last 100 years, the working classes in Europe and North America enjoyed a relatively brief period of integration and stability during the middle decades of the twentieth century. During those years, they most effectively challenged the economic and political domination of capital. In the sense that a working class in flux has come back since 1973, we can say that social relations have undergone a change. But this change is a return to a historically more typical condition.

Spatially the fragmentation of the working class and the successful defense by upper-income groups of their privileges is reflected in increasing social segregation. This segregation and its resultant defensive communities have in turn, of course, reinforced social fragmentation. While these tendencies have been exacerbated by economic contraction and fiscal crisis, their beginnings are in American suburbanization and European peripheral housing development for the working class and, even earlier, in the establishment of workers' districts and upper-class residential areas within compact cities. Again we conclude that we are witnessing a long-term, ongoing trend.

THE STATE

More than 100 years ago, Marx commented on how "the original isolation of the separate nationalities is destroyed by the developed mode of production and intercourse and the division of labor between various nations naturally brought forth by these" (Marx, 1970: 58). The mutual dependence of national economies has a long history, and the economic fate of most nations has largely depended on decisions made outside their boundaries. Nevertheless, within the core countries, national governments expanded in power after Marx's time; by the 1930s, they had become heavily involved in the provision of social benefits, using more and more sophisticated modes of intervention so as to achieve much greater control over internal economic processes and outcomes. The intensified internationalization of both finance and production, as well as the movement of firms and financial establishments to tax havens and regulation-free zones, that has occurred in the last 20 years appears to have reduced that capacity for control.

Likewise, at the subnational level, the seemingly increased mobility of capital has pushed regions and municipalities into heightened competition with each other and created a consequent inability to tax and regulate industries within their borders because of the potential for corporate blackmail. The separation of production from consumption functions and their location in different political jurisdictions as well as the isolation of low-income populations have reduced the capability of regional and local governments to intervene to achieve economic stability and social welfare.

Both of these tendencies, however, must be seen as long-term, contradictory, and subject to cyclical variation. While national governments have lost control over their economies, international coordination—through both regional groups, like the European Economic Community, and broader ones, like the International Monetary Fund and the Organization for Economic Cooperation and Development—has increased, allowing new modes of regulation. Moreover, declines in support for the welfare state and Keynesian economic management have by no means meant their elimination nor total lack of effectiveness. Even in small states that are highly dependent on world markets, national governments have been able to affect the economic course of their societies (Katzenstein, 1985). Competition among states for resources and markets is certainly nothing new in the history of capitalism, nor is it more intense now than at the rise of the German and American industrial empires.

Similarly, at the subnational level, the competition for new development that today drives regional governments into financing private enterprises and city governments into public-private partnerships has many precedents:

> The role of the growth machine as a driving force in U.S. urban development has long been a factor in U.S. history, and is nowhere more clearly documented than in the histories of eighteenth- and nineteenth-century American cities. . . . The creators of towns and the builders of cities strained to use all the resources at their disposal, including crude political clout, to make great fortunes out of place. . . .
>
> These communities competed to attract federal land offices, colleges and academies, or installations such as arsenals and prisons as a means of stimulating development. . . . The other important arena of competition was also dependent on government decision making and funding: the development of a transportation infrastructure that would give a locality

better access to raw materials and markets [Logan and Molotch, 1987: 52-53].

Logan and Molotch (1987: 55) go on to describe the success of Los Angeles business interests in securing infrastructure support, allowing it to become the preeminent city of the West, even though "it had none of the 'natural' features that are thought to support urban growth: no centrality, no harbor, no transportation crossroads, not even a water supply."

Spatial fragmentation has simultaneously insulated governments and business elites from working-class demands and heightened pressures from communities organized around issues of collective consumption (Castells, 1983). The mid-1970s were an important break point, both in the United States and in Europe, in marking the quiescence of urban social movements. To some extent this decline resulted from structural economic change (see S. Fainstein, 1987); but more significant was the mediating function played by ideological factors that undercut the subjective bases for mobilization (Smith and Judd, 1984). Whether or not urban social movements again become important political factors influencing state policy remains an open question. Current spatial trends allow both possibilities: on one hand, the constraints on autonomous local state action limit the possible rewards available to local movements; on the other, the increased concentration of wealth in city centers causing them to be more vulnerable to threat, the growing homogeneity of neighborhoods, and the further drop in living standards for various strata of the population provide the basis for new efforts. Community groups are currently on the defensive, but their situation does not preclude future mobilization and is not sharply different from that of the early 1960s.

CONCLUSIONS

Our conclusion then is that the current period has indeed seen the victories for capital delineated by the theorists of restructuring. These victories, however, are neither cause nor consequence of a radically new set of spatial relations or technological capabilities. Rather, we have witnessed a century-long weakening of technical constraints on the movement of capital and people, which has allowed capitalists, through the strategies of individual firms and the activities of government, to continue to make use of space as an instrument in the process of creative destruction and worker control. In an earlier period, Marx predicted that the proximity of workers in the factory would produce solidarity. But except for a brief moment in its history, capitalism has been characterized far more strongly by centrifugal impulses than by consolidating ones. If space is to prove contradictory in its effect and provide a vehicle for mass mobilization, it will be through the resurgence of urban social movements. Their current quiescence in the core states of Europe and North America, however, indicates that their development is not a necessary outcome of spatial fragmentation and collective consumption.

We should say in closing that we do not deny that significant changes of two kinds have characterized world capitalism in the postwar years, with an accelerating tendency since the early 1970s. First, on the grandest scale, the balance among capitalist nation-states has altered substantially as the core has moved increasingly toward the Pacific basin. Second, Third World peripheries have been increasingly integrated into the world capitalist economy, with great consequences for their internal patterns of spatial development. Most evidently, developmental patterns have been marked by rural depopulation, hyperurbanization, the simultaneous explosion of informal economies, and formal exploitation of an emergent urban proletariat producing for the world economy. While we cannot discuss the spatial impacts of these changes within the confines of this chapter, we can note once again that shifts in the capitalist core, market penetration of rural hinterlands, and extraordinary outbursts of urbanization have always been the essential dynamics of the modern world capitalist economy. A line of similar processes can be traced back from Third World cities of the 1980s to Tokyo in the 1920s, Berlin and Chicago in 1880, and London in 1800. Our argument is not that of the modernization theorists—that each country will proceed through a set of similar stages. Rather we contend that the processes of capitalist incorporation and exploitation follow a typical dynamic within the space economy, one that can be traced back to at least the nineteenth century. The specific outcomes of these processes are contingent on the interaction of world economic forces and particular local factors.

Unquestionably, the post-1973 capitalist crisis of stagnation and declining profits has been subjected to a "spatial fix" as part of a broad effort at restructuring. But the succession of periods of stability, crisis, and restructuring are endemic to the dynamics of capitalist evolution, and the effort by capital to exploit territorial divisions so as to lower factor costs and divide the working class represents a perpetual theme of capitalist development. The jeopardy to the success of the present effort comes from the state's loss of control over the processes of production and reproduction that are no longer contained within its boundaries. While this situation advantages multinational corporations by freeing them from much regulation and taxation, it undercuts capitalist stability by keeping the state from acting on their behalf to prevent overproduction and ensure popular tranquility. The spatial fix might thus be temporary and, like previous similar efforts, merely a precursor to the next crisis.

NOTES

1. Markusen et al. (1986) argue that, in general although not always, mature industries are dominated by large oligopolistic firms that decentralize units of their operation. Massey (1984) is less willing to portray a predictable cycle of geographical behavior but establishes a typology of concentrated versus disaggregated spatial structures of particular firms. For these theorists, the behavior of the firm is tied to its particular profit situation rather than to the recent general restructuring of capitalism. To the extent, however, that most firms simultaneously experienced a profit squeeze and contracting markets, one could predict widespread synchronous efforts either to resort to a deconcentration strategy so as to reduce costs or to close down facilities altogether. Markusen et al.'s formulation also implies that nations primarily dependent on mature oligopolistic industries will experience more deconcentration than countries harboring young innovative industries.

Both Markusen and Massey focus on manufacturing and administrative control of manufacturing; they do not treat the geography of "back-office" functions, nor do they examine nonmanufacturing firms. But there is little reason to think that the separation of routine office functions such as billing, direct mail, inventory control, data processing, and so on from headquarters activities is a markedly different phenomenon from the earlier separation of product manufacturing, although, as will be discussed below, its impact may be significant.

2. Direct foreign investment in manufacturing enterprises in Third World countries for reexport (i.e., platform economies) is the one factor with important spatial consequences that apparently constitutes a genuinely new phenomenon. It is, however, a logical outgrowth of the increasing decentralization of production that had previously occurred within the advanced countries.

3. The reduction in time that women devote to nurturing, as a result both of lower fertility rates and of work outside the home, indicates a dramatic change in the relations of reproduction and the character of family life. In the United States, average household size declined from about 4.5 persons in 1920 to 3.4 persons in 1950, and finally to 2.7 persons in 1984 (U.S. Bureau of the Census, 1985a: table 54). Elderly family members live in their own, geographically separated households or in a variety of institutional settings. The welfare state serves both the old and the young, replacing the functions of the family and deconstructing the household.

REFERENCES

BLUESTONE, B. and B. HARRISON (1986) "The great American jobs machine." Prepared for the U.S. Congress, Joint Economic Committee, December.

CASTELLS, M. (1983) The City and the Grass Roots. Berkeley: University of California Press.

CASTELLS, M. (1984) "Towards the informational city? High technology, economic change, and spatial structure: some exploratory hypotheses." Working Paper No. 430. Berkeley: University of California, Institute of Urban and Regional Development.

CASTELLS, M. (1985) "High technology, economic restructuring, and the urban-regional process in the United States," pp. 11-40 in M. Castells (ed.) High Technology, Space, and Society. Beverly Hills, CA: Sage.

EHRENREICH, B. (1983) The Hearts of Men. New York: Doubleday.

FAINSTEIN, N. (1987) "The underclass/mismatch hypothesis as an explanation for black economic deprivation." Politics and Society, 15 (4): 403-451.

FAINSTEIN, S. (1987) "Local politics and economic discontent," pp. 323-342 in M. P. Smith and J. R. Feagin (eds.) The Capitalist City. Oxford: Basil Blackwell.

FAINSTEIN, S. and N. I. FAINSTEIN (1986) "Regime strategies, communal resistance, and economic forces." Chap. 7 in S. Fainstein et al. (eds.) Restructuring the City. New York: Longman.

FEAGIN, J. R. and M. P. SMITH (1987) "Cities and the new international division of labor: an overview," pp. 3-34 in M. P. Smith and J. R. Feagin (eds.) The Capitalist City. Oxford: Basil Blackwell.

FRIEDMANN, J. (1986) "The world city hypothesis." Development and Change 17: 69-83.

FROBEL, F., J. HEINRICHS, and O. KREYE (1980) The New International Division of Labor. Cambridge: Cambridge University Press.

GERSHUNY, J. I. (1978) After Industrial Society. London: Macmillan.

GORDON, D. M. (1978) "Capitalist development and the history of American cities," pp. 25-63 in W. K. Tabb and L. Sawers (eds.) Marxism and the Metropolis. New York: Oxford University Press.

HARVEY, D. (1973) Social Justice and the City. Baltimore: Johns Hopkins University Press.

HARVEY, D. (1985) Consciousness and the Urban Experience. Baltimore: Johns Hopkins University Press.

HARVEY, D. (1986) The Urbanization of Capital. Baltimore: Johns Hopkins University Press.

HARVEY, D. (1987a) "The geographical and geopolitical consequences of the transition from Fordist to flexible accumulation." Presented at the Conference on America's New Economic Geography, Rutgers University, Center for Urban Policy Research, Washington, DC, April.

HARVEY, D. (1987b) "Flexible accumulation through urbanization: reflections on 'post-modernism' in the American city." Presented at the 6th Urban Change and Conflict Conference, University of Kent, Canterbury, Kent, U.K., September.

HILL, R. C. (1985) "The global factory and the company town: the changing division of labor in the international automobile industry." Prepared for a conference sponsored by the ISA, the Department of Sociology, and the Centre for Urban Studies and Urban Planning, University of Hong Kong, August 14-20.

JAMESON, F. (1984) "Postmodernism, or the cultural logic of late capitalism." New Left Review 146: 53-92.

KATZENSTEIN, P. (1985) "Small nations in an open international economy: the converging balance of state and society in Switzerland and Austria," pp. 227-256 in P. B. Evans et al. (eds.) Bringing the State Back in. New York: Cambridge University Press.

KATZNELSON, I. (1981) City Trenches: New York: Pantheon.

KONRAD, G. and I. SZELENYI (1977) "Social conflicts of underurbanization," pp. 157-173 in M. Harloe (ed.) Captive Cities. New York: John Wiley.

LAWRENCE, R. Z. (1984) "Sectoral shifts and the size of the middle class." Brookings Review (Fall).

LIEPETZ, A. (1986) "New tendencies in the international division of labor: regimes of accumulation and modes of regulation," pp. 16-40 in A. J. Scott and M. Storper (eds.) Production, Work, and Territory. Boston: Allen & Unwin.

LOGAN, J. R. and H. L. MOLOTCH (1987) Urban Fortunes. Berkeley: University of California Press.

MANDEL, E. (1975) Late Capitalism. London: New Left Books.

MARKUSEN, A. (1985) Profit Cycles, Oligopoly, and Regional Development. Cambridge: MIT Press.

MARKUSEN, A., P. HALL, and A. GLASSMEIER (1986) High Tech America. Boston: Allen & Unwin.

MARX, K. (1970) The German Ideology. New York: International.

MASSEY, D. (1984) Spatial Divisions of Labor. London: Macmillan.

McKELVEY, B. (1963) The Urbanization of America, 1860-1915. New Brunswick, NJ: Rutgers University Press.

MINGIONE, E. (1981) Social Conflict and the City. Oxford: Basil Blackwell.

MINGIONE, E. (1983) "Informalization, restructuring and the survival strategies of the working class." International Journal of Urban and Regional Research 7 (3): 311-339.

MOSS, M. (1986) "Telecommunications and the future of cities." Land Development Studies 3: 33-44.

NELSON, K. (1986) "Labor demand, labor supply and the suburbanization of low-wage office work," pp. 149-171 in A. J. Scott and M. Storper (eds.) Production, Work, Territory. Boston: Allen & Unwin.

New York Affairs (1987) Special Issue: "Ethnic neighborhoods old and new." Winter, 10 (1).

NOYELLE, T. J. and T. M. STANBACK, Jr. (1983) The Economic Transformation of American Cities. Totowa, NJ: Rowman & Allanheld.

O'CONNOR, J. (1973) The Fiscal Crisis of the State. New York: St. Martin.

PAHL, R. E. (1980) "Employment, work and the domestic division of labor." International Journal of Urban and Regional Research 4 (1): 1-20.

PAHL, R. E. (1984) Divisions of Labor. London: Basil Blackwell.

PERRY, D. C. (1987) "The politics of dependency in deindustrializing America: the case of Buffalo, New York," pp. 113-137 in M. P. Smith and J. R. Feagin (eds.) The Capitalist City. Oxford: Basil Blackwell.

PIORE, M. (1979) Birds of Passage. New York: Cambridge University Press.

SASSEN-KOOB, S. (1986) "New York City: economic restructuring and immigration." Development and Change 17: 85-119.

SASSEN-KOOB, S. (1987) The Mobility of Labor and Capital. New York: Cambridge University Press.

SAYER, A. (1986) "Industrial location on a world scale: the case of the semiconductor industry," pp. 107-124 in A. J. Scott and M. Storper (eds.) Production, Work, Territory. Boston: Allen & Unwin.

SCOTT, A. J. and M. STORPER (1986) "Industrial change and territorial organization: a summing up," pp. 303-312 in A. J. Scott and M. Storper (eds.) Production, Work, Territory. Boston: Allen & Unwin.

SMITH, M. P. and D. R. JUDD (1984). "American cities: the production of ideology," pp. 173-196 in M. P. Smith (ed.) Cities in Transformation. Beverly Hills, CA: Sage.

SMITH, M. P. and R. TARDANICO (1987) "Urban theory reconsidered: production, reproduction and collective action," pp. 87-110 in M. P. Smith and J. R. Feagin (eds.) The Capitalist City. Oxford: Basil Blackwell.

SMITH, N. (1984) Uneven Development. Oxford: Basil Blackwell.

STORPER, M. and A. J. SCOTT (1986) "Production, work, territory: contemporary realities and theoretical tasks," pp. 3-15 in A. J. Scott and M. Storper (eds.) Production, Work, Territory. Boston: Allen & Unwin.

SZELENYI, I. (1981) "Structural changes of and alternatives to capitalist development in the contemporary urban and regional system." International Journal of Urban and Regional Research 5 (1): 1-15.

U.S. Bureau of the Census (1960) Historical Statistics of the United States, Colonial Times to 1957. Washington, DC: Government Printing Office.

U.S. Bureau of the Census (1985a) Statistical Abstract of the United States: 1986. Washington, DC: Government Printing Office.

U.S. Bureau of the Census (1985b) Current Population Reports. Series P-60, no. 146. Washington, DC: Government Printing Office.

VOGEL, D. (1986) "The future of New York City as a financial center: an international perspective." Presented at the Workshop on Metropolitan Dominance, Committee on New York City, Social Science Research Council, New York, November.

VOGEL, L. (1983) Marxism and the Oppression of Women: Toward a Unitary Theory. New Brunswick, NJ: Rutgers University Press.

WALDINGER, R. D. (1986) Through the Eye of the Needle: Immigrants and Enterprise in New York's Garment Trades. New York: New York University Press.

WALLERSTEIN, I. (1980) The Modern World System II. New York: Academic Press.

ZARETSKY, E. (1976) Capitalism, the Family, and Personal Life. New York: Harper & Row.

2

Pittsburgh in Transition: Consolidation of Prosperity in an Era of Economic Restructuring

GORDON L. CLARK

IT IS EASY TO SAY that the economy of the Pittsburgh region is in transition. All the statistics indicate a regional economy very different from what it was ten years ago, and all the indications are that Pittsburgh will be vastly different in another ten years. For many of those living in Pittsburgh, the word *transition* is too antiseptic a word to describe the changes of the past decade. For unemployed steelworkers, middle-level managers, and some corporate executives, the word *transition* is facile. They have experienced drastic declines in their livelihoods and standards of living. Their jobs, prospects for employment, and relatively high incomes have apparently evaporated for good. A combination of local and national industrial economic stagnation, relocation of production, and competition from overseas imports have dealt many local residents a massive blow to their well-being.[1]

From 1979 to 1986, the regional economy lost approximately 55,000 steel industry jobs, and a further 50,000 manufacturing jobs. There is little chance of these jobs returning to the Pittsburgh area, despite wishful thinking and planning by some community activists. Based on the geographical patterns of restructuring in the U.S. steel industry, in 1989 the local steel industry will employ less than one-quarter of what it did in 1979, the most recent peak year of employment in the industry (see Clark, 1988, for a detailed assessment of restructuring in the steel

AUTHOR'S NOTE: *Thanks to Brian Berry, Ben Fischer, and Joel Tarr for comments and advice on previous drafts of this chapter. They should not be held responsible for any errors or omissions, or the opinions expressed herein.*

industry). Only in the past year has employment in the Pittsburgh region regained the level of employment characteristic of the last years of the 1970s. For displaced workers, perhaps the most appropriate word to describe their experiences during the 1980s would be *disaster*.[2]

The notion of transition implies change from one condition to another. At the simplest level, it is clear that the Pittsburgh region is changing from a production economy to a service economy. As we will see, Pittsburgh shares many of the same employment growth trends that have been observed in other similar-sized cities around the country (Stanback and Noyelle, 1982). Of course, the city of Pittsburgh has always had a significant services component, principally business services associated with corporate headquarters. But if we look more closely at this part of the local economy, it will become apparent that even this sector is in transition. Pittsburgh is not the corporate headquarters city it once was, even though its fortunes are more than ever linked to the service economy.[3]

The problem is that the simple "service" image of economic transition hides more than it reveals. What is lacking in such a broad characterization is a sense of the impact that the new service economy will have on the Pittsburgh region. The service economy of Pittsburgh will be quite different than the corporate service economy: lower-paid jobs, unstable employment, and few opportunities for advancement. While it is clear that this type of service economy is expanding in Pittsburgh, in relation to the past, the transition may have relatively limited benefits for the local economy as a whole.

In this essay, two arguments are made about the recent past. First, economic restructuring has had quite distinct spatial effects. The mill towns and cities of the surrounding river valleys have borne the brunt of economic decline in the steel and related manufacturing industries. In this respect, the city of Pittsburgh has consolidated its economic dominance of the region. Second, the corporate headquarters sector is now less important to the city of Pittsburgh, in terms of both its absolute numbers of employment and its relative contribution to total city employment. Thus, unlike many other large U.S. manufacturing cities, economic restructuring has been both spatially dispersed and highly localized. The central city has not experienced wholesale economic destruction, but its relative place in the American corporate hierarchy has declined.[4]

Two arguments are made about the future of the Pittsburgh region. First, having lost the steel industry, the mill towns will just fade away, leaving the metropolitan core of the region. Second, the city will face increasing competition from its suburbs, as the significance of the corporate sector further declines, and as newer firms and industries locate outside the urban core. In this respect, the outskirts of the region are in danger of becoming museum pieces, dominated by the shadows of derelict mills and abandoned communities. The city itself is also in danger of becoming a museum piece, dominated by a massive architecture of former corporate power, buildings of the past but not of the future. Basically, Pittsburgh will become more like American cities of a similar size. Its position in the history of American industrialization will not be challenged; but its history will be a thin veneer on a fairly ordinary city.

A BRIEF HISTORICAL PERSPECTIVE

In the preface of her book on the history of Pittsburgh from 1681 to 1906, Sarah Killikelly (1906: iii) made two observations that have served as common reference points in understanding the history of the region. First, "the history of Pittsburgh is essentially the history of Allegheny County—the greater Pittsburgh" and "the city, as the metropolis of the county and the vast busy region adjacent thereto, has stood out in the affairs of the State and nation, and the history of the world, as a leader in the evolution of the industrial arts for nearly a century." She went on to say that "for this reason it has been impossible to confine the treatment [her history] to Pittsburgh proper, or to give distinct place to any lesser city or town within the district." While Pittsburgh, as a city, has been the geographical focus of the region, its significance as an industrial center is, more often than not, linked to the extended network of industrial towns and cities that straggle up and down the adjoining river valleys.[5]

The second observation made by Killikelly concerned the significance of industry to the region. As noted, she believed the region to have extraordinary status in the "history of the world," being one of the principal industrial leaders of the world. Given her own timeframe, the turn of this century, it should be apparent from our perspective in time that Pittsburgh has been at the center of industrial production and technology for nearly 200 years. The significance of this history for Killikelly was more profound than this obvious acknowledgment would seem to imply. She wrote that "the chief factor in its [Allegheny County and the city of Pittsburgh] rise and progress has been manufacturing.

All others have been secondary." The very character of the city, according to Killikelly, was set by the industrial fabric of the region. In this sense, her history of Pittsburgh was the history of the region and its industry, and how these essential elements permeated local institutions like schools, hospitals, and the city government.[6]

The notion of Pittsburgh as a specialized and spatially extended industrial region has dominated our imagination for many years. For those of us whose job it is to describe American cities, Pittsburgh has been an easy case; heavy industry and river valleys dominate our stories. For industrial geography textbook writers, it has been used as the paradigmatic case of how locational advantage structures and differentiates the landscape.[7]

E. Willard Miller's (1962) textbook is a good example of this perspective. In describing the industrial geography of Anglo-America, he began with a broad set of specialized industrial regions, and then described each region in terms of its particular geography and industry. In this setting, Pittsburgh is described by Miller as a manufacturing region of cities extending from Cleveland through to the city of Pittsburgh. The physical geography of the region dominates Miller's description. The river valleys, their relation to the surrounding hills and mountains, and the availability of raw materials all figure prominently in the story.

According to Miller, the development of the Pittsburgh region owed its origins to local high-grade coal reserves. In the earliest of times, Pittsburgh had a prosperous iron industry. With the introduction of the Bessemer furnace (which needed very good quality coking coal) in the mid-1860s, the region prospered. In fact, the southwestern region of Pennsylvania became the heartland of American heavy industry. Even after these reserves were depleted, the river routes in and out of the region made importation of needed coal quite feasible. With technological leadership, available raw materials, and a highly developed industrial system, Pittsburgh was able to translate its initial advantages into long-run economic growth.

The river valleys were a resource and a means of transportation. Water was used for cooling in steel mills and furnaces, and the rivers were used to transport both raw materials and finished products. Moreover, the river valleys mapped out the organization of commerce and production. Production along the river valleys was quite literally designed in a sequential manner, taking the materials produced at one town up and down the rivers. In this manner, the river valleys were extended integrated production complexes, with specific towns and cities specialized by function and task. Given the narrowness of the valley floors, there was little opportunity to spread production back into the surrounding country. Each town along the valleys had a certain role or roles in the overall geography of production.

Of course, Pittsburgh has become more than steel mills and heavy manufacturing. Since World War II, the Pittsburgh regional economy has diversified and developed in other sectors, consistent with patterns in the U.S. economy at large (see Stanback and Noyelle, 1982, for more details). So, for example, if we consider Table 2.1, it should be apparent that durable manufacturing's share of total nonagricultural employment in the Pittsburgh region (SMSA) has fallen precipitously since 1954. It once claimed over 36% of all employment. By 1979, its share had declined to about 22%. The overall share of employment due to manufacturing has also dramatically declined. In 1954, manufacturing in the region accounted for about 41% of employment. By 1979, this share had declined to just 27%. Not only did manufacturing's relative share of employment decline, this sector also had significant losses in absolute levels of employment. So, by 1979, firms in the manufacturing sector employed 20% fewer people than they did in 1954.

Despite losses of about 80,000 jobs in the manufacturing, mining, and transportation sectors, nonagricultural employment grew by about 22% during this same period (approximately 175,000 jobs). Employment in the trade and services; finance, insurance, and real estate (FIRE); and government sectors dramatically increased. In Table 2.1, it is shown, for example, that employment in the service sector grew by 110%. Moreover, relative shares of total employment also increased dramatically with, for example, the service sector's share of total employment rising from 12% to about 21% of total nonagricultural employment. While these sectors did not dominate the regional economy as much as they did the national economy (compare Stanback and Noyelle, 1982), their share of employment is indicative of a more general economic transformation that has occurred in the United States and in Pittsburgh.[8]

If Table 2.1 is examined carefully, it is evident that these trends were affected by economic shifts and changes. Each economic downturn brought lower levels of employment, and in each economic upturn, peak levels of employment were successively less than the previous peak.[9] For example, the peak year of employment in durable manufacturing in the 1970s (1979) was a year in which the employment level was 100,000

TABLE 2.1
Employment by Sector for Pittsburgh SMSA, 1954-1979 (in thousands)

	Year[a]								
Sector	*1954*	*1956*	*1962*	*1969*	*1971*	*1979*	*Percentage of Change 1954-1979*	*Percentage of Total 1954*	*Percentage of Total 1979*
Total nonagriculture	776.6	847.6	732.5	877.9	840.9	948.3	22.1		
Mining	14.8	14.3	8.4	7.5	3.5	10.9	–26.3	1.9	1.1
Construction	37.1	43.5	34.2	44.0	43.8	50.8	36.9	4.8	5.3
Manufacturing	318.3	354.1	260.2	293.0	249.3	253.1	–20.5	41.0	27.0
Durable goods	269.0	303.7	217.1	245.7	202.2	209.0	–22.3	34.6	22.0
Nondurable goods	49.3	50.4	43.1	47.3	47.1	44.1	–10.5	6.3	4.6
Transportation and public utilities	69.8	72.2	55.6	60.2	57.6	58.3	–16.4	8.9	6.1
Trade	150.1	159.1	145.3	174.8	177.6	213.5	42.3	19.3	22.5
FIRE	28.6	30.5	31.9	36.8	38.0	44.6	55.9	3.7	4.7
Services	94.3	105.6	118.2	154.1	157.9	198.1	110.0	12.1	21.0
Government	63.6	68.3	78.7	107.5	111.1	119.0	86.3	8.3	12.5

SOURCE: U.S. Department of Labor, unpublished metropolitan (SMSA-adjusted) employment and earnings file.
a. Years selected correspond to the sequence of business cycle fluctuations, beginning with 1954 (trough) and ending with 1979 (peak), in local economic activity. Note that employment figures for each year are not yearly averages. Rather, end-of-year figures are used.

lower than the peak year (1956) in the 1950s. In contrast, even in recession, the service sector gained, especially in overall shares of total employment. While there were marked cyclical patterns in some industries, the process of transformation was relatively slow (compared to other manufacturing centers in the United States). The Pittsburgh region went from a predominately manufacturing economy to a predominately service economy in the space of 25 years. In doing so, the local economy was able to expand by creating new jobs in newer sectors, thereby replacing employment losses in the traditional sectors.

Despite the sectoral shifts of the post-World War II era, the older production landscape still dominates the towns and cities of the valleys, and perhaps even people's conception of the present. Mile after mile of abandoned steel mills, fabricating plants, and all kinds of heavy metals manufacturing plants are a testament to the power of past locational imperatives like terrain and transportation. One might quibble with the intellectual power of Miller's analysis, given current developments in location theory, which focus more upon the social relations of production than local physical geography (compare with Massey, 1984; Warde, 1985). And, as we shall see, these kinds of imperatives are of less relevance when we come to understanding the Pittsburgh of tomorrow. Still, we should be wary of belittling the significance of traditional location factors. The inherited landscape of the region is still important in how displaced workers understand their predicament.

NATIONAL RECESSION/ LOCAL ECONOMIC CRISIS

What of the Pittsburgh region today? It should be apparent to any visitor to the region that a lot has happened since 1979. At the beginning of the chapter, I noted that the region lost over 100,000 steel and manufacturing jobs between 1979 and 1986. In this period, the national economy went through one of the worst recessions since the Great Depression of the 1930s. Recovery has advanced but has been limited to certain economic sectors. For the Pittsburgh region, the national recession was more like economic disaster: massive job loss, extraordinarily high levels of unemployment, and permanent closing of many manufacturing plants. The economic recovery has been very slow in coming to Pittsburgh, and has not significantly benefited those who lost their jobs for good.[10]

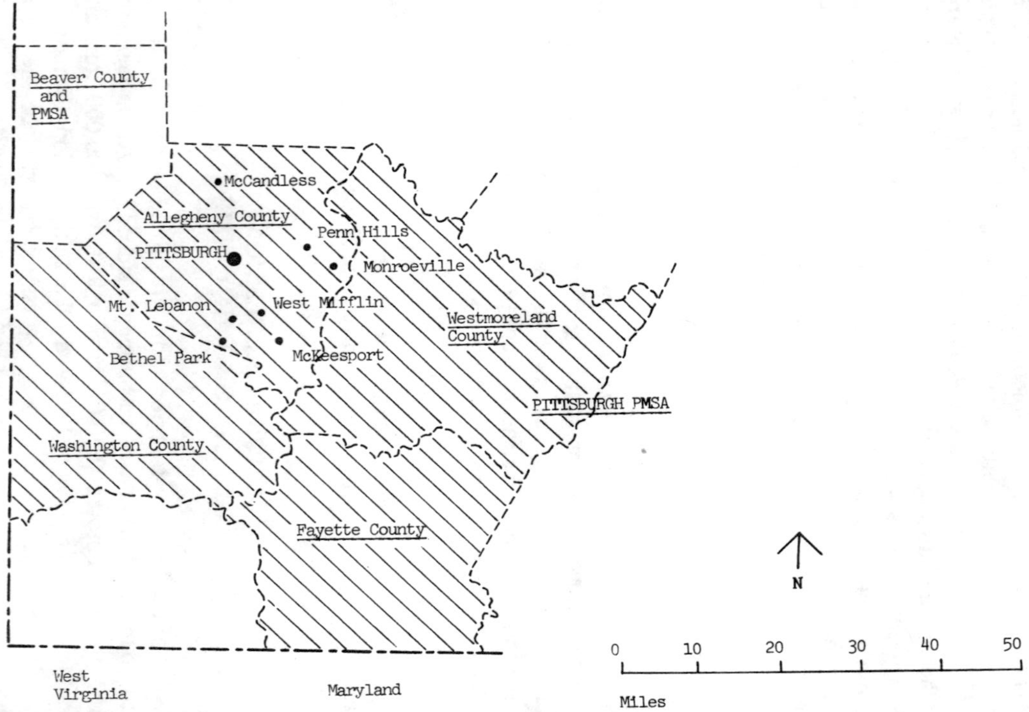

Figure 2.1 The Pittsburgh Region

TABLE 2.2

Employment Levels and Change for Selected Years by Sector: Beaver (B) and Pittsburgh (P) PMSAs, 1975-1984 (in thousands)

Sector	*1975*		*1979*		*Percentage of Change 1975-1979*		*Percentage of Total 1979*		*1980*		*1984*		*Percentage of Change 1980-1984*		*Percentage of Total 1984*	
	B	*P*	*B*	*P*	*B*	*P*	*B*	*P*	*B*	*P*	*B*	*P*	*B*	*P*	*B*	*P*
Nonagricultural employment	75.9	853.3	80.4	916.0	5.9	7.3			76.4	909.7	55.7	836.0	–27.1	–8.1		
Manufacturing	39.2	219.8	38.7	224.5	–1.2	2.1	48.1	24.5	35.3	213.5	17.2	147.6	–51.3	–30.8	30.8	17.7
durable	36.2	177.5	35.6	180.7	–1.6	1.8	44.2	19.7	32.3	171.1	14.8	109.8	–54.2	–35.8	26.5	13.1
nondurable	3.1	42.3	3.1	43.8	0.0	3.5	3.8	4.8	3.0	42.5	3.2	37.8	6.7	–11.1	5.7	4.5
Nonmanufacturing	36.6	631.8	41.5	688.7	13.3	9.0	51.6	75.2	41.0	694.8	38.5	687.6	–6.1	–1.0	69.1	82.2
construction	3.5	39.0	4.3	46.4	22.8	18.9	5.3	5.1	3.5	44.0	2.8	37.5	–20.0	–14.7	5.0	4.5
transportation, public utilities	4.6	55.3	5.7	58.8	24.0	6.3	7.0	6.4	5.7	54.4	5.2	49.0	–8.8	–9.9	9.3	5.8
trade	11.1	187.1	12.3	203.5	10.8	8.8	15.3	22.2	12.5	206.8	11.4	206.7	–8.8	0.0	20.4	24.7
FIRE	1.6	40.5	1.8	44.7	12.5	10.3	2.2	4.9	2.0	45.7	1.9	49.1	–5.0	7.4	3.4	5.8
services and mining	7.9	183.4	9.1	218.8	15.1	19.3	11.3	23.7	9.5	222.9	10.0	238.3	–5.2	6.9	17.9	28.5
government	7.9	126.5	8.1	116.5	25.3	–7.9	10.1	12.7	7.8	121.1	7.1	106.0	–8.9	–12.5	12.7	12.7

SOURCE: Commonwealth of Pennsylvania.

To understand Pittsburgh today, we must consider the ramifications of the recent local economic crisis for both the towns and the cities of the region and the region's industrial structure. (See Figure 2.1.) To do so requires that we look first at the overall dimensions of local job loss.

In Table 2.2, employment by sector for the period 1975 to 1984 is reported for two Pennsylvania Metropolitan Statistical Areas (PMSA), Beaver and Pittsburgh. The first PMSA encompasses Beaver County. The second PMSA encompasses Allegheny, Fayette, Washington, and Westmoreland counties. The Beaver PMSA is downstream on the Ohio River from the City of Pittsburgh and was dominated by the steel industry. Previously, it was included as a county in the Pittsburgh SMSA. Fayette County, to the southeast of Pittsburgh, was previously in another SMSA, but has now been incorporated into the newly formed Pittsburgh PMSA. This county has been historically less dependent upon manufacturing. Thus the previously reported SMSA data are not strictly comparable. To understand the magnitude of changes since 1979, data were collected for the previous four years and the subsequent four years.

Overall, in the period 1975 to 1979, both PMSAs recorded modest gains in total nonagricultural employment; 5.9% and 7.3% gains, respectively. During the period 1980 to 1984, both PMSAs recorded dramatic declines in total employment of -27.1% and -8.1%, respectively. The Beaver PMSA was far more affected than the Pittsburgh PMSA. Indeed, if we look more closely at Table 2.2, the Beaver PMSA recorded declines in almost all sectors during this period, whereas job loss in the Pittsburgh PMSA was limited to manufacturing. Furthermore, in those sectors where both PMSAs lost employment, job loss was catastrophic in the Beaver PMSA compared to the Pittsburgh PMSA. So, for example, the Beaver PMSA lost 51.3% of its manufacturing jobs during this period compared to Pittsburgh, which lost 30.8% of its manufacturing jobs. Both figures are startlingly large, given the thousands of people involved. These losses were so significant that, by 1984, manufacturing was "only" 30.8% of total employment in the Beaver PMSA—compared to 48.1% in 1979, and 17.7% of total employment in the Pittsburgh PMSA—compared to 24.5% in 1979.

The scale of job loss, its speed, and the economic hardships visited upon permanently displaced workers shocked industry and civic leaders. Even though there had been a modest decline in manufacturing employment in the Beaver PMSA during the previous four years (-1.2%), there was no conception in the local community that such a catastrophe was possible. Although unemployment statistics are somewhat unreliable, especially at larger values, the scale of this economic disaster should be readily apparent to the reader. Nationally, unemployment rose to about 11% around the trough of the recession (late 1982). In the Pittsburgh region, it was not uncommon during this period to see unemployment rates that led the nation. For example, Beaver County recorded an annual rate of unemployment in 1983 of 21%.[11]

Likewise, if we were to consider PMSA-county employment patterns, similar patterns would emerge, although with a couple of qualifications. There were significant job losses in all counties during the 1980 to 1984 period, with Beaver County leading the decline. But more than massive job loss occurred in this period, especially if we take into account the whole ten-year period. With respect to the changing fortunes of the three southern counties of the Pittsburgh PMSA, job loss in these counties was quite modest compared to Beaver County, and there was even net job growth during the total period 1975 to 1984, despite the recession of 1982. Although more recent data are not officially available, these counties have further consolidated their growth, and even Allegheny County has recovered to at least its 1975 employment level. Beaver County, though, has further declined.[12]

When we remember that, not too many years ago, manufacturing was the heart of this region, these job losses point to a fundamental collapse of the region's traditional economy and a substantial spatial-economic reorientation of the region. It should be quite obvious that the source of overall employment growth in the three southern Pittsburgh-PMSA counties was in the nonmanufacturing sectors, specifically in services, FIRE, and trade (see Table 2.3). But even more important to recognize is the implied spatial/sectoral restructuring of the region. Employment in the older industrial towns of the river valleys basically collapsed during the last five years. These towns have become quite literally boarded-up ghost towns, dying in the shadows of their decaying factories and mills. On the other hand, service-oriented employment in the growing southern and eastern towns, away from the river valleys, has expanded. As well, service employment in the city of Pittsburgh has also expanded, both downtown and around the eastern inner-city areas adjacent to the various universities and hospitals.

Documenting these emerging spatial and economic patterns is quite difficult, given the available data. There is very little in the way of industry data that describe the region below the county scale. Nevertheless, one indication of the changing geography of prosperity of the

TABLE 2.3
Annual Employment and Unemployment by Minor Civil Division for the Pittsburgh PMSA, 1975-1985[a]

	Employment (in thousands)				*Percentage of Change*	*Percentage of Change*	*Unemployment (in percentages)*					
County/City/Town	*1975*	*1979*	*1980*	*1984*	*1975-1979*	*1980-1984*	*1975*	*1977*	*1979*	*1981*	*1983*	*1985*
Allegheny												
Bethel Park	13.2	15.1	14.4	13.5	14.4	–6.2	4.9	4.7	3.9	4.6	8.7	7.1
McCandless	9.5	11.3	10.8	10.2	18.9	–5.5	3.5	3.6	3.1	3.5	6.8	3.9
McKeesport	11.4	10.9	12.8	11.7	–4.4	–8.5	9.5	10.1	9.4	18.5	18.8	10.9
Monroeville	12.1	14.1	12.8	12.1	16.5	–5.5	4.9	4.9	4.2	7.3	10.4	7.8
Mt. Lebanon	14.1	15.0	14.2	13.3	6.4	–6.3	3.7	3.5	2.9	3.4	6.1	5.0
Penn Hills	23.1	25.0	23.8	22.2	8.2	–6.7	6.1	6.6	6.1	6.4	11.9	8.4
Pittsburgh	170.3	168.5	175.2	161.7	–1.1	–7.7	8.0	8.2	7.3	7.4	12.1	9.1
Plum	8.8	10.7	10.5	10.0	21.6	–4.8	5.3	5.3	4.5	6.1	12.9	8.6
Ross	13.6	15.9	14.5	13.7	16.9	–5.5	4.1	4.1	3.6	4.7	9.6	5.7
Shaler	12.8	14.4	13.9	13.0	12.5	–6.5	4.2	4.4	3.9	5.0	9.4	7.1
West Mifflin	10.0	10.7	10.9	10.1	7.0	–7.3	6.5	6.6	5.9	8.0	15.3	11.1
Westmoreland												
Hempfield	16.5	19.0	17.9	19.3	15.1	7.8	6.2	6.0	5.7	7.5	14.1	10.0
North Huntingdon	11.6	13.1	13.0	12.2	12.9	–6.1	6.1	6.1	5.9	8.0	16.0	11.4

SOURCE: Commonwealth of Pennsylvania.
a. Based on residential data for towns and cities with greater than 25,000 population.

region is provided in Table 2.3. Previously, it was noted that employment in the three southern and eastern counties of the Pittsburgh PMSA grew quite substantially during the 1975 to 1985 period. The few data we have for these counties at the town and city scale, specifically Westmoreland County, reinforce this conclusion. Despite quite high rates of unemployment in the early 1980s, employment in the towns of Hemfield and North Huntington grew significantly during the total 1975 to 1985 period, with Hemfield recording the only net employment gain of any of the towns listed in the table in both subperiods.

The established south-hills middle- and upper-class communities of Allegheny County, Mt. Lebanon and Bethel Park, and the established eastern townships of the county, Monroeville and Penn Hills, maintained their overall levels of employment for the ten-year period, even though they all lost employment in the 1980 to 1985 period. Comparatively speaking (compare with the working-class town of McKeesport), these communities were not as affected by the recession of 1982 as other communities—witness the lower rates of unemployment. As well, the newer north-hills communities expanded during the ten-year period, with McCandless maintaining employment and unemployment levels, and the other close communities of Ross and Shaler advancing their shares of overall employment. Note that two of the older valley communities of McKeesport and West Miflin recorded higher rates of unemployment (though not like the Beaver PMSA), and little overall growth in employment. Indeed, like the city of Pittsburgh, McKeesport recorded employment declines in both subperiods of the 1975 to 1985 period and higher than average unemployment rates during the entire ten years.

During the past ten years (and especially the past five years), the economy of the Pittsburgh region has become more dependent upon the service sectors and has lost a large part of its industrial base. In terms of history, the Pittsburgh region is no longer an industrial economy. While Killikelly's (1906) depiction of Pittsburgh as the leader of industrialization has a great deal to say about the history of the region, this image is now a historical legacy, not a lived reality. Over just five years, the economy of the region has been fundamentally altered.

As a consequence, and in conjunction with the development of the service sectors of the region, industrial communities like Aliquippa and Homestead located in the river valleys have become shells of the past, while the newer suburban communities have grown, linked to quite different industries. In this respect, the region is dominated by two fundamentally different economic landscapes. The landscape of the past is the industrial landscape that dominates the river valleys. While obvious enough within the city of Pittsburgh, especially in the mill neighborhoods of Braddock and Hazelwood, its worst effects are in the southwest and northwest sections of Allegheny County and in the Beaver PMSA.[13] Abandoned mills and derelict factories are stark reminders of massive worker displacement and unemployment. The other landscape is the office landscape, which is readily apparent in downtown Pittsburgh but finds its newest expression in the outer suburbs and communities of the eastern and southern counties of the Pittsburgh PMSA. Offices, industrial parks, hospitals, and shopping malls are the dominant architectural representations of this economic mode.

PITTSBURGH AS A SERVICE ECONOMY

Clearly, there have been many changes in the spatial configuration of the economy of the region. But, to understand Pittsburgh today, and its possible futures, we should look closer at the economy of the region. In particular, we need answers for the following questions. What are the implications for local incomes of having lost so many manufacturing jobs during the last five years? Will the Pittsburgh region thrive on the expanding service sector jobs, as it once did on manufacturing jobs? What are the prospects for the region in terms of the growth sectors of the local economy? To understand the concrete implications of these questions, we again compare data from two of the regions' counties, Beaver and Allegheny.

In Table 2.4, average county employment and wages are compared for Beaver and Allegheny Counties. Unfortunately, these 1985 data are the latest data available. More than likely, the manufacturing sectors' employment and wages are overestimated in both counties—in the years since the third quarter of 1984, more manufacturing jobs have been lost in the region, and wages have been steadily falling in related sectors. Even so, these figures do serve to highlight the relative differences between the counties, and between the sectors in terms of their wage and employment patterns. In particular, the manufacturing sector in both counties was still important for both areas in terms of employment. The importance of manufacturing goes beyond jobs, however. In terms of income, the manufacturing sector provides some of the highest monthly

TABLE 2.4
Employment and Average Monthly Wages by Industry/Sector for Third Quarter 1984: Beaver County and Allegheny County

Industry[a]	Employment Allegheny	Employment Beaver	Average Monthly Wages ($) Allegheny	Average Monthly Wages ($) Beaver
Manufacturing	101,222	17,516	2320	2184
durable	73,048	15,125	2417	2247
nondurable	28,174	2,391	2066	1785
Construction	31,421	2,872	1911	2235
Transportation	14,793	1,444	1773	1300
Public utilities	16,563	2,618	2313	2437
Retail trade	54,501	5,559	1018	809
Food service	41,469	3,218	493	456
Banking	15,369	626	1804	1165
Insurance	11,120	386	1687	1279
Health services	59,076	3,745	1687	1567
Legal services	4,991	152	2046	1656
Local government	45,363	5,004	1588	1504
State government	8,101	785	1561	1510
Federal government	15,010	435	2202	1954

SOURCE: Commonwealth of Pennsylvania (1985c), Beaver and Pittsburgh PMSAs.
a. Selected industries; employment figures do not add to total county employment.

incomes in the region. Moreover, the spread in manufacturing incomes between the counties is very small ($2320 for Allegheny and $2184 for Beaver), especially when we compare counties in terms of their average monthly incomes derived from the banking sector ($1804 for Allegheny and $1165 for Beaver).

Based on Table 2.4, I wish to make three related observations. First, as employment in the manufacturing sectors has declined, per capita incomes have also declined as the local labor force has come to depend more upon the service economy for their livelihoods. Whether we consider Allegheny or Beaver Counties, the wage pattern between sectors is such that service sectors typically pay less each month than manufacturing sectors. This is most pronounced for the traditional customer service sectors, like retail, banking, and food service. In these sectors, average monthly wages are about one-half, one-quarter, and one-half, respectively, of manufacturing wages. Of course, these sectors are dominated by part-time work, so we should not automatically conclude that wages in these sectors represent individuals' total monthly incomes. Given that jobs in these three sectors are, in total, more than all manufacturing jobs in both counties, one should be wary of underemphasizing their importance, especially considering the dramatic job loss in manufacturing in recent years. Because many of the workers in these sectors are female, these part-time jobs have had an important role in maintaining family incomes during the past few years.

Second, there are a set of public service sectors with comparatively high average monthly wages. Specifically, these sectors include local, state, and federal governments, and public utilities. These sectors are very important for both counties in terms of their contribution to total employment. In Allegheny County, these four sectors contribute about as many jobs as manufacturing. Beaver County, however, is not as fortunate, given the relatively high wages of these sectors and the likely further decline in local manufacturing jobs. As well, other, more specialized sectors, like legal and health services, have quite high average monthly incomes.

There is a further pattern embedded in Table 2.4. Although both counties have significant numbers of these more specialized higher-paying sectors, Allegheny County benefits more than Beaver County in this regard. Thus the third observation about patterns of employment and wages in the regional economy has to do with the spatial concentration of these sectors in the city of Pittsburgh as opposed to the outlying river communities. Insurance, health services, and legal services are very important to Allegheny County, both for total employment and for average monthly wages. In the third quarter of 1984, these three sectors in total provided more jobs than durable manufacturing. This was not the case by any means in Beaver County. As well, the average monthly wages in these sectors in Allegheny County were consistently higher than the average monthly wages of these same sectors in Beaver County. In fact, in two of the three sectors, average monthly wages in Allegheny County were $400 more than in Beaver County.

These patterns are also evident in the occupational profiles of the two areas. Table 2.5 summarizes by broad categories the occupational mix for the Beaver PMSA and the Pittsburgh PMSA. Note that, as was previously observed, Beaver County and the Beaver PMSA are the same geographical area. Allegheny County, however, is just one county (albeit the dominant county) of the Pittsburgh PMSA. Thus Tables 2.4 and 2.5 are not strictly comparable. Still, going on the previous data, the patterns evident in Table 2.5 are consistent with Table 2.4. In particular,

TABLE 2.5
Occupational Profiles for Beaver and Pittsburgh PMSAs
Compared to Pennsylvania, 1984

Occupation	*Percentage of PMSA*[a] *Beaver*	*Percentage of PMSA*[a] *Pittsburgh*	*PMSA/PA Ratio* *Beaver*	*PMSA/PA Ratio* *Pittsburgh*
Executive/managerial	6.0	9.1	0.7	1.1
Professional	9.1	12.3	0.8	1.1
Technical	2.5	3.3	0.8	1.1
Sales	8.5	10.2	0.9	1.1
Administrative support	14.5	17.1	0.8	1.0
Service	12.4	13.2	1.0	1.1
Farming	0.7	0.7	0.4	0.4
Precision production	15.7	13.5	1.2	1.0
Machine operators	14.5	8.6	1.2	0.7
Transportation	7.2	5.4	1.4	1.0
Handlers/cleaning	8.2	6.2	1.4	1.0

SOURCE: Commonwealth of Pennsylvania (1985c), Beaver and Pittsburgh PMSAs.
a. Due to the fact that not all occupational categories are presented in this table, columns do not add to 100%.

the Pittsburgh PMSA has many more—proportionally and in absolute numbers—higher-level management occupations than the Beaver PMSA. On the other end of the scale, the Beaver PMSA has proportionally more production-related occupations than the Pittsburgh PMSA. Of course, given the relative sizes of the Beaver and Pittsburgh PMSAs, the absolute numbers of production-related occupation employees would be much higher in Pittsburgh than in Beaver.

Finally, sales, service, and administrative occupations are more important in the Pittsburgh PMSA than in the Beaver PMSA. Overall, compared to the Pennsylvania economy, the Pittsburgh PMSA appears slightly more specialized in the higher-management occupations than most areas of Pennsylvania. Certainly, this is the case when compared to the Beaver PMSA. Even so, we must be cautious of these figures. In the last few years, major corporations located in the Pittsburgh PMSA have laid off significant numbers of middle-management staff. Thus it is more than possible that Table 2.5 overestimates the relative significance of managerial occupations in the Pittsburgh PMSA.

It has been observed in other contexts that the service economy can be a mixed blessing (see Stanback and Noyelle, 1982; Bluestone and Harrison, 1986). While there are a number of quite high-paying jobs, there are many service jobs that have very low rates of pay, and limited opportunities for full-time employment. As well, it is thought that the trend of increasing wage inequality is related to an increasingly bimodal wage distribution in the service sector. These patterns may be attributed in part to the relative lack of unionization in the service sectors and consequent limited job tenure in these sectors. Whatever the reasons for these patterns, there can be no doubt as to the significance of the service economy for the United States—in terms of jobs, income, and newer occupations. Still, there is a strong suspicion that the emerging service economy is more polarized than the traditional manufacturing economy.[14]

These kinds of patterns are replicated in the Pittsburgh regional economy. Of course, there can be no doubt that the service economy is very important to the counties and PMSAs of the region, and to the region as a whole. In an era of massive job retrenchment, the viability of the service sector has saved many families from starvation. Even so, on average, service jobs in Beaver County pay less than the same service jobs in Allegheny County. Moreover, Beaver County seems to have proportionately fewer of the higher-paying service jobs than Allegheny County. And there are proportionally fewer higher-paid occupations in Beaver County than in the Pittsburgh PMSA. Essentially, the river valley communities most affected by retrenchment in their manufacturing sectors are least endowed with diverse, high-paying job opportunities in the service sectors. It appears that the service economy is stratified *spatially* and *occupationally*. In this respect, there may be a quite distinct spatial hierarchy to the service economy, locally and nationally (Stanback and Noyelle, 1982).

PITTSBURGH IN FIVE YEARS' TIME

Forecasting the future is fraught with all kinds of dangers. When we look to the future, we tend to extrapolate dominant themes and dominant presumptions. That is, forecasting takes the past and present and extends them into the future. In doing so, we assume that our understanding of the present, framed by our past experience, is an adequate datum point for knowing the future. And yet, it is just as plain to anyone who considers recent history that there have been dramatic surprises, events that were not considered possible or were considered irrelevant to the overall path of the regional economy. So, for example, the collapse of much of Pittsburgh's manufacturing industry through the early years of this decade caught many commentators by surprise.

Going on the evidence of the late 1970s, a period of extraordinary production in the manufacturing sector, one might have reasonably expected continued levels of prosperity, if not employment growth.

Those who previously argued that Pittsburgh would be less dependent upon manufacturing in the future than in the past were surprised at the speed of economic restructuring and worker displacement. Indeed, one of the dominant themes that emerges from interviews with displaced workers in the valleys around Pittsburgh is the level of surprise at their predicament. In the past, local industry has gone through successive periods of economic prosperity, decline, and rejuvenation. Among those workers whose lives, and the lives of their families, were framed by work in the mills, many experienced extended periods of layoff and then recall. In the most recent downturn, what began as a yet another round of retrenchment became a fundamental reconfiguration of the local economy. For many workers, it has taken a number of years to realize the fundamental nature of this restructuring. Even now there are a significant number of workers and citizens who believe that Pittsburgh will again be a major center for manufacturing.

In five years' time, I expect that Pittsburgh will be both less than what it is now, certainly much less than what it was ten years ago, and more than what it is now. The balance of growth and decline will probably favor growth. Of course, this scenario presumes no further surprises and a continuing erosion of the traditional economic base of the region. Most crudely, I expect that manufacturing jobs will continue to shrink, both absolutely and in terms of their share of overall regional employment. I also expect that service employment will continue to expand, albeit slowly and in the lower end of the occupational scale. Essentially, the Pittsburgh region will have more people working than ever before, in rather different occupations than traditionally associated with Pittsburgh, and at lower average rates of pay. As a consequence, while the local economy may not shrink in terms of available real household spending power, real personal incomes will likely further decline.

In Table 2.6, the 30 largest employers of the region are ranked by their employment levels at the end of 1985. There are three striking patterns in this table. First, the top 10 employers are split quite dramatically between those who are losing employees (concentrated in the top group of 5 employers), and those who are gaining employees (concentrated in the next group of 5 employers). In the former group, as in many areas, the government sector dominates. But unlike many other areas,

TABLE 2.6
Largest Employers in the Pittsburgh Region, 1984-1985[a]

Ranking, Name	*End of 1985*	*End of 1984*	*Percentage of Change*
1. Local government	77,018	77,500	−0.6
2. Westinghouse	23,149	23,845	−2.9
3. Federal government	17,174	17,400	−1.3
4. State government	13,526	14,700	−7.9
5. United States Steel Corporation	10,621	10,639	−0.2
6. University of Pittsburgh	8,379	7,839	+6.9
7. Mellon Bank	7,800	7,500	+4.0
8. USAir	7,100	6,058	+17.2
9. Giant Eagle (supermarket)	6,457	6,000	+7.6
10. The May Co. (Kaufmann's)	5,955	5,945	+0.2
11. Duquesne Light	4,685	4,741	−1.2
12. J. C. Penney (includes Thrift Drug)	4,503	4,400	+2.3
13. Allegheny Ludlum	4,500	4,489	+0.3
14. Schneider Group of Companies	4,450	3,850	+15.6
15. PPG Industries	4,190	4,270	−1.9
16. Bell of Pennsylvania	4,182	4,474	−6.5
17. Pittsburgh National Bank	3,644	3,525	+3.4
18. McDonald's Corporation	3,518	3,100	+13.5
19. Allegheny General Hospital	3,405	3,517	−3.2
20. Sears, Roebuck and Company	3,400	3,400	0
21. Alcoa	2,748	2,692	+2.1
22. St. Francis Medical Center	2,741	2,825	−2.9
23. Volkswagen	2,700	2,700	0
24. Eat'N Park Restaurants	2,676	2,645	+1.2
25. Carnegie-Mellon University	2,661	2,425	+9.7
26. AT&T	2,644	2,490	+6.2
27. LTV Steel	2,500	4,100	−39.0
28. Associated Dry Goods (Horne's)	2,408	2,127	+13.2
29. Presbyterian-University Hospital	2,378	2,124	+11.9
30. Western Pennsylvania Hospital	2,350	2,600	−9.6

SOURCE: *Pittsburgh Post-Gazette* (Tuesday, January 26, 1986); Commonwealth of Pennsylvania.
a. Includes part-time employees.

represented in this group are two of the older major industrial firms of the region—Westinghouse (the largest private employer in Pennsylvania) and United States Steel Corporation (the firm which owes its origins to Andrew Carnegie). In contrast, the next group of employers is essentially service oriented—ranging from the university (University of Pittsburgh) to the bank (Mellon) to the air carrier (USAir) to the retail stores (Giant Eagle and Kaufmann's). Going on the wage figures introduced in Table 2.5, the first five employers tend to pay higher wages than the second group of employers. These same groups of employers are, respectively, losing employees and gaining employees.

A second observation has to do with the importance lower down the scale of service employers, ranging from fast-food firms like McDonalds to utilities like Bell of Pennsylvania and to hospitals like Allegheny General. In this respect, the rank of local employers in terms of their numbers of employees correlates with previously introduced data regarding sectoral and occupational trends in the local region. A third observation has to do with the patterns of growth and decline among the firms represented in Table 2.6. The hospital sector is growing very rapidly (although this is somewhat mixed) and the banking and university sectors are also significantly contributing to expanding employment opportunities. In terms of decline, the major industrial conglomerates are slightly, and in a couple of instances rapidly, shedding employment.

In the next five years, one could imagine that government employment will continue to decline, as will industrial conglomerate employment, and there may be slight increases in service sector employment. But these increases will be slight. The universities are reaching plateaus in their employment because of declining student numbers, doubts about financing education, and some (University of Pittsburgh) cutbacks in staffing. Conventional retail stores are under considerable financial stress, especially the older established stores, and the utilities that depend upon the base population of the area for their employment levels are also likely to see limited growth. There may be further increases in hospital employment, although this will be mixed between the different hospital systems depending upon their levels of dependence on government medicare/medicaid payments.

If we now consider the likely patterns of labor demand by occupation, we should expect that those occupations tied to the service sector will be most in demand. The U.S. Department of Labor and the Commonwealth of Pennsylvania have constructed estimates of labor demand by occupation for the Beaver and Pittsburgh PMSAs for the period 1980 to 1990. Whatever the likely real figures for each occupation at the end of this decade, it is apparent that projected occupational demand mirrors previously identified trends toward a lower-level local service economy.

Very few of the occupations likely to grow during the next few years are professional in the sense that they require college education, and most require just high school education. Most growing occupations have high rates of turnover, and pay much less than the traditional blue-collar job.

What about the much ballyhooed high-tech revolution, and its Pittsburgh representatives in the local universities and high-tech parks? Unpublished employment figures from the Commonwealth of Pennsylvania suggest that high-tech employment (that is, employment directly attributable to innovative technology) at the beginning of 1985 numbered about 4,000 to 5,000. While there may well be significant gains in employment in high-tech-related sectors, the actual impact on the Pittsburgh economy is going to be very limited. The reality is that these sectors employ very few workers, and although employment growth may be startling, the base is very small.[15]

In five years' time, the Pittsburgh economy will be much like it is today. There will not be a high-tech employment miracle; people will likely work in service and managerial jobs similar to those they work in today, and those still working in manufacturing will likely be facing further layoffs. In five years though, the occupational and sector mix of service jobs will likely tend toward lower-paying jobs. In this scenario, restructuring will further consolidate the prosperity of the city of Pittsburgh, its southeastern suburban communities, and areas around the airport transportation corridor. The traditional production communities of the river valleys will be further impoverished and denuded of viable employment opportunities.

CONCLUSIONS

By my reckoning, I doubt that Pittsburgh will again have such an important share of national industry. But, in asserting this claim, I do not mean to suggest that Pittsburgh will lose all of its manufacturing, nor do I mean to suggest that Pittsburgh will altogether lose its historic role in the management and planning of industry. Given the historic connections of some of the major manufacturing conglomerates to the region, we might reasonably claim that Pittsburgh will retain significant management functions. This role is fragile, as the experience of towns and cities caught up in the recent wave of corporate mergers shows. We need not go beyond Pittsburgh for recent evidence of this process. When Gulf Oil was merged with Standard Oil of California, Pittsburgh lost many of Gulf Oil's former employees to California, and the professional employment base itself. In this context, the (re)organization of major industrial corporations like USX, Rockwell, and Westinghouse will have far-reaching implications for the long-run wage and employment structure of the region.

If we look toward the distant future—say 2006, the centenary of the publication of Sarah Killikelly's book—I doubt that Pittsburgh will be described as she described it so graphically. Instead of being described as the "furnace" of America, the major actor of the American industrial revolution, Pittsburgh will probably be described by what it was and what it became. That is, in terms of its past: what it was, its historical legacy, and the residue physical landscape—its buildings and monuments of a previous era. I doubt that many mills will be left, save perhaps one or two that will be preserved as historical parks. The landscape will be more pleasant, certainly less industrial, more in keeping with many other similar-sized cities around America. Pittsburgh also will be described in terms of its late-twentieth-century transformation; from an internationally oriented industrial city to a regionally oriented service economy.

In a sense, this scenario is rather optimistic. It supposes a spatial and economic transformation quite in keeping with the experiences of many other American cities through much of this century. Indeed, one might claim that Pittsburgh is one of the last nineteenth-century industrial cities to go through this transformation. And I have no doubt that, looked at historically, the locational imperatives that were so important in shaping the local economy are now of limited utility in describing the current spatial logic of industrial production, in relation both to other American and to Third World cities.

Having noted these broad trends, we must not be overly sanguine about the enormous personal costs involved in this transformation.[16] We cannot ignore the plight of displaced workers, either through an oblivious description of patterns, or an indifference born of an abstract idealization of the processes of economic adjustment (compare with Lawrence and Litan, 1985). The reality is that many workers and their families are caught between a dangerous whirlpool of economic change (Charybdis) and government policy (Scylla) that seem quite indifferent to their plight. In this respect, the government appears as the mythical monster, stirring the whirlpool through its budgetary and trade policies, and doing little to alleviate the welfare consequences of its policies. It remains to be seen what the cost of the restructuring of Pittsburgh will be in terms of human suffering.

NOTES

1. See Mieszkowski (1985) for a study of the regional impacts of foreign competition, Grunwald and Flamm (1985) for a series of studies on the globalization of production, and Berry et al. (1985) for a detailed study of the implications of economic restructuring for American industry and cities. Note that there is some debate over the relative contributions of foreign competition, declining domestic demand, and low labor productivity to the collapse of American heavy industry, especially steel.

2. In a study commissioned by the United Steelworkers of America, Locker/Albrecht Associates, Inc. (1985: 1) began by asserting three conclusions: (1) The industry is in deep trouble, (2) some major companies face immediate crisis, and (3) long-run prospects for recovery are dim. Without concerted government, union, and company cooperation, the whole industry, and its dependent communities, face a fundamental threat to their survival.

3. The chief economist of the Mellon bank noted some time ago that, without the service economy, the area economy would have literally collapsed (quoted in the *Pittsburgh Business Times-Journal*, March 31, 1986: 6). While there are reasons to be cautious about the substantive importance of the service economy in providing long-term employment, there can be no doubt as to the significance of this sector during the past five or so years.

4. In recent years, there has been a good deal of debate over how regional economic change should be analyzed. A variety of methods have been promoted; see Clark et al. (1986) on the technical aspects of regional economic dynamics, Markusen (1985) for a series of case studies, and Massey and Meegan (1985) on alternative approaches to industrial geography.

5. See also the edited volume by Lorant (1964) for a wealth of detail on Pittsburgh as a region and as a city.

6. See Lubove (1969) and Courvares (1984) for examples of histories of Pittsburgh's institutions framed with reference to industry and economy.

7. For example, David Houston (1986: 1) began his study of the history of the Pittsburgh economy by remarking that Pittsburgh "burned like a roman candle on the American industrial scene" and "represented an intensification of this process (uneven development of capitalist cities)." Houston associates the rise of Pittsburgh with the emergence of American industrial capitalism ("the raw and triumphant power of capitalism").

8. Berry and Elster (1985) have made similar observations regarding the relative specialization of the local economy, in regard both to national trends and to local historical trends.

9. This pattern of successive cyclically based declines in employment is typical of the experience of many cities in the United States (see Clark et al. 1986: chaps. 4-6) and overseas (see, for example, Lloyd and Shutt, 1985, on recession and regional restructuring in the British economy). Essentially, recession enables firms permanently to lay off labor in the guise of a temporary layoff.

10. A study by Management Science Resources Inc. noted that laid-off workers have fared poorly in obtaining local jobs, even with retraining. The demand for labor is now quite different than it was 20 years ago, being split between professional employment and low-skilled service jobs that typically hire women. According to the study, many unemployed workers will only find employment if they move (quoted in the *Pittsburgh Post-Gazette*, June 5, 1986: business page).

11. Underreporting of unemployment is a severe problem in the region. In a 1985 study prepared for the Diocesan Task Force on Unemployment, it was reported that the "true" unemployment rate for the Pittsburgh region (six counties including Lawrence, Butler, Beaver, Allegheny, Washington, and Greene) was 17.0% as opposed to 9.5% as reported by the federal government (quoted in *Labor Day, 1985*: 1, available from the Catholic Diocese). A more scholarly survey of local unemployment for the Youngstown/Warren Ohio SMSA reported that the federal Bureau of Labor Statistics (BLS) underestimated local unemployment by as much as 15 percentage points. For example, BLS estimated Youngstown unemployment to be 15.2% as of June 1984, whereas the Current Population Survey estimate of local unemployment was 29.3%. Similar levels of differences were found for other areas around Youngstown (Buss, 1986).

12. In fact, the Commonwealth of Pennsylvania estimated that Allegheny County gained some 30,000 jobs during the year 1985. While not as blessed with rapid job growth, Fayette, Washington, and Westmoreland counties also experienced significant net job additions during the year (2000, 3000, and 5000 jobs, respectively).

13. See Tarr and Di Pasquale (1982) on Hazelwood as a mill town in the city. Their study was one of only a few to consider a contemporary urban industrial neighborhood in relation to broader trends in the region and economy.

14. Berry and Elster (1985) argued that bipolarization of earnings and employment in Pittsburgh is consistent with national trends. They, like me, suggested that this bipolarization has two dimensions: an occupational dimension and a spatial dimension. High-income professionals will be further concentrated in the far suburbs (despite gentrification; see Berry, 1985), and low-income service employees will occupy (occupationally and spatially) inner-city locations. Granted, there is a continuing debate about the generality of this conclusion. Nevertheless, the implications of restructuring in Pittsburgh point to this kind of pattern.

15. It has been noted that the high-tech sector has been more concerned in the past couple of years with capital financing than hiring labor. Indeed, the average size of high-tech firms in the region appears to be less than the nation. This view is in stark contrast with the media representation of the Pittsburgh region. For example, the *Dallas Morning News* (September 9: 1) led off with an article extolling the "new" Pittsburgh—an "economic dawn founded upon education and medicine." The practical realities are something quite different.

16. For a recent attempt to estimate the wage and income effects of worker displacement in the Pittsburgh region, see Jacobson (1986).

REFERENCES

BERRY, B.J.L. (1985) "Islands of renewal in seas of decay," pp. 69-96 in P. Petersen (ed.) The New Urban Reality. Washington, DC: Brookings Institution.

BERRY, B.J.L. and S. ELSTER (1985) "What lies ahead for urban America?" pp. 17-36 in J. R. Hitchcock and A. McMaster (eds.) The Metropolis: Proceedings of a Conference in Honour of Hans Blumenfeld. Toronto: University of Toronto, Department of Geography.

BERRY, B.J.L., S. SANDERSON et al. (1985) "Growth and adjustment in an economy undergoing structural transformation." Pittsburgh, PA: Carnegie-Mellon University, School of Urban and Public Affairs.

BLUESTONE, B. and B. HARRISON (1986) "The great American job machine: the proliferation of low wage employment in the U.S. economy." Prepared for the Joint Economic Committee, U.S. Congress, Washington, DC.

BUSS, T. (1986) "Assessing the accuracy of BLS local unemployment rates: a case study." Industrial and Labor Relations Review 39: 241-250.

CLARK, G. L. (1988) "Corporate restructuring in the steel industry: adjustment strategies and local labor relations," in J. Hughes and G. Sternlieb (eds.) America's New Economic Geography. New Brunswick, NJ: Rutgers University, Center for Urban Policy and Research.

CLARK, G. L., M. GERTLER, and J. WHITEMAN (1986) Regional Dynamics: Studies in Adjustment Theory. Hemel Hempstead and Boston: Allen & Unwin.

Commonwealth of Pennsylvania (1983) Statewide and Major Labor Market Areas Trends and Outlook for Nonagricultural Wage and Salary Workers, 1980 and Projected 1990. Harrisburg, PA: Office of Employment Security.

Commonwealth of Pennsylvania (1984) Pittsburgh SMSA Occupational Trends and Outlook for Total Civilian Employment 1980 and Projected 1990. Harrisburg, PA: Office of Employment Security.

Commonwealth of Pennsylvania (1984-1985) JTPA Service Delivery Areas Occupational Trends and Outlook for Total Civilian Employment 1980 and Projected 1990. Harrisburg, PA: Office of Employment Security.

Commonwealth of Pennsylvania (1985a) Pennsylvania Civilian Labor Force Series by Minor Civil Division. Harrisburg, PA: Office of Employment Security.

Commonwealth of Pennsylvania (1985b) Beaver County Annual Planning Information Report, Fiscal Year 1986. Harrisburg, PA: Office of Employment Security.

Commonwealth of Pennsylvania (1985c) Pittsburgh PMSA Annual Planning Information Report, Fiscal Year 1986. Harrisburg, PA: Office of Employment Security.

COURVARES, F. G. (1984) The Remaking of Pittsburgh: Class and Culture in an Industrializing City 1877-1919. Albany: State University Press of New York.

GRUNWALD, J. and K. FLAMM [eds.] (1985) The Global Factory: Foreign Assembly and International Trade. Washington, DC: Brookings Institution.

HARRISON, B., C. TILLY, and B. BLUESTONE (1986) "The great U-turn: increasing inequality in wage and salary income in the US." Cambridge: MIT, Department of Urban Studies and Planning. (mimeo)

HOUSTON, D. B. (1986) "A brief history of the process of capital accumulation in Pittsburgh: a Marxist interpretation." Pittsburgh, PA: University of Pittsburgh, Department of Economics. (mimeo)

JACOBSON, L. (1986) "Labor mobility and structural change in Pittsburgh 1977-82." Kalamazoo, MI: W. E. Upjohn Institute for Employment Research. (mimeo)

KILLIKELLY, S. H. (1906) The History of Pittsburgh: Its Rise and Progress. Pittsburgh: B. C. & Gordon Montgomery Co.

LAWRENCE, R. Z. and R. E. LITAN (1985) "Living with the trade deficit: adjustment strategies to preserve free trade." Brookings Review 4 (1): 3-13.

LLOYD, P. and J. SHUTT (1985) "Recession and restructuring in the north-west region 1974-82: the implications of recent events," pp. 64-87 in D. Massey and R. Meegan (eds.) Politics and Method: Contrasting Studies in Industrial Geography. New York: Methuen.

Locker/Albrecht Associates, Inc. (1985) Confronting the Crisis: The Challenge for Labor. Report to United Steelworkers of America, New York.

LORANT, S. [ed.] (1964) Pittsburgh: The Story of an American City. Lenox, MA: Authors Edition, Inc.

LUBOVE, R. (1969) Twentieth Century Pittsburgh: Government, Business, and Environmental Change. New York: John Wiley.

MARKUSEN, A. (1985) Profit Cycles, Oligopoly, and Regional Development. Cambridge: MIT Press.

MASSEY, D. (1984) Spatial Divisions of Labor. London: Macmillan.

MASSEY, D. and R. MEEGAN [eds.] (1985) Politics and Method: Contrasting Studies in Industrial Geography. New York: Methuen.

MIESZKOWSKI, P. (1985) "The differential effect of the foreign trade deficit in regions of the United States," pp. 346-363 in J. M. Quigley and D. L. Rubinfeld (eds.) American Domestic Priorities: An Economic Appraisal. Berkeley: University of California Press.

MILLER, E. W. (1962) The Geography of Manufacturing. Englewood Cliffs, NJ: Prentice-Hall.

STANBACK, T. and T. NOYELLE (1982) Cities in Transition: Changing Job Structures in Atlanta, Denver, Buffalo, Phoenix, Columbus (Ohio), Nashville, and Charlotte. Totowa, NJ: Allanheld, Osmun.

TARR, J. and W. C. LAMPERES (1981) "Changing fuel use behavior and energy transitions: the Pittsburgh smoke control movement, 1940-1950." Journal of Social History 14: 561-588.

TARR, J. and D. DI PASQUALE (1982) "The mill town in the industrial city: Pittsburgh's Hazelwood." Urbanism Past and Present 7 (13): 1-14.

WARDE, A. (1985) "Spatial change, politics and the division of labor," in D. Gregory and J. Urry (eds.) Social Relations and Spatial Structures. London: Macmillan.

3

New Trends in the Sociospatial Organization of the New York City Economy

SASKIA SASSEN

THIS CHAPTER FOCUSES on two issues. One is the spatial expression of major changes in the economic base of large cities. The other is whether and how spatial arrangements in the sphere of social reproduction, in turn, have a feedback effect on the spatial organization of production. The empirical elaboration of the argument uses the case of New York City, and includes the results of two current research projects. The concern is not an inventory of land uses one can detect in such cities. The concern is rather with examining the links between key changes in the spatial organization of the city and key economic sectors. This link can operate directly through the sphere of production and indirectly through that of social reproduction. This link can be very evident, as in the expansion of the luxury office complex into areas of the city hitherto considered inadequate for these economic sectors—notably the West Side of midtown Manhattan. Or this link can be rather difficult to recognize, as is the case with the proliferation of small firms, many of them informal, resulting from changes in the consumption structure brought about by the transformation in the economic core of such cities.

There are three processes of economic and spatial organization I see as central to the question addressed in this chapter. One is the expansion and consolidation of the producer services and international corporate

AUTHOR'S NOTE: *I want to thank the Social Science Research Council for supporting the preparation of an earlier version of this chapter, presented to the Workshop on the Dual City, Social Science Research Council, New York City, October 29, 1986.*

headquarters sector into the economic core of major postindustrial cities. While this sector may not account for the majority of jobs, it establishes a new regime of economic activity and the associated spatial and social transformations evident in these cities. A second process is the downgrading of the manufacturing sector, a notion I use to describe a mode of political and technical reorganization of manufacturing that is to be distinguished from the decline and obsolescence of manufacturing activities; according to such a notion, then, the downgraded manufacturing sector represents a mode of incorporation into the "postindustrial" economy. The third process is the informalization of a growing array of economic activities, a process that encompasses certain components of the downgraded manufacturing sector; informalization represents a mode of reorganizing the production and distribution of goods and services that falls outside or escapes the regulatory apparatus of the formal economy.

While there is some overlap between downgrading and informalization, each of these three processes can be seen as separate and distinct modes of economic organization and the corresponding uses of space: the postindustrial city of luxury high-rise office and residential buildings located largely in Manhattan; the old dying industrial city of low-rise buildings and family-type houses, located largely in the outer boroughs; and the Third World city imported via immigration and located in dense groupings spread all over the city. Each of these can be seen to contain distinct income-occupational structures and concomitant residential and consumption patterns, represented in the expansion of a new urban gentry alongside expanding immigrant communities. One key question is whether they are indeed as separate as they appear, each belonging to a different spatiotemporal configuration. Or can we identify a common dynamic that generates segmentation in the organization of economic activity so that growth in one feeds growth in the other. What presents itself as disarticulation can then be seen, in this second case, as a form of articulation that makes possible the reproduction of a politically, ethnically, and spatially differentiated organization of the economy.

There are three sets of questions I will pursue in examining each of these processes. First, what are the sources of growth of each? Second, and partly overlapping with the first question, what are the linkages among these three processes? Third, how do these processes create and destroy employment opportunities while also constituting adequate labor supplies?

In the first section, I briefly describe key issues and questions emerging out of my research. These are, clearly, just a few of all the possible questions that the evidence on New York City warrants; but they are a beginning. In the following three sections, I discuss the characteristics and implications of each of the above processes.

A CONCEPTUAL PERSPECTIVE

Major changes in the technical and spatial organization of the economy at the local, national, and global levels have fed the demand for the kinds of services and centralized control operations that major postindustrial cities produce. Here the notion of a "world city" (Friedmann and Wolff, 1982), or what I have elsewhere called "global cities," becomes of great use. The need for nodal points to coordinate global economic activity contributes to the emergence of world cities. To this coordinating role I add that of sites for the *production* of a large array of inputs and "organizational commodities" necessary for global control and coordination (Sassen-Koob, 1984a).

These inputs need to be produced. We can think of the producer services sector as a sort of new basic industry and of cities like New York, Los Angeles, or London as ideal sites of production for this industry. A concern with production also makes it possible to shift the focus away from the large multinational corporations and onto the vast array of small firms involved in the production of such services. Furthermore, it leads to an investigation of the organization of this industry: its patterns of subcontracting, suppliers, networks, input and output markets, and other components. That is to say, while large multinationals are the key economic actors in today's national and world economy, there is a vast array of economic activities that are necessary for the type of economic system such corporations engender, but that take place outside these corporations. One aspect of the significance of major cities in the current economic phase lies in this role they play as sites for the location of this new basic industry, rather than as sites for the location of corporate headquarters.

A question that emerges out of this concerns the consequences of such a new economic core in major cities for the organization of other economic sectors. Specifically, does the existence of a dynamic growth sector feed the expansion of what appear to be declining or backward economic sectors, such as the downgraded manufacturing and informal sectors? What are the linkages? To what extent do the most sophisticated white-collar industries that characterize the new economic core need access to a broad range of industrial services located in easy access, notably in Manhattan, though these services typically lack the bidding power to locate within easy access? We see here one of many contradictions built into the spatial organization of the city's economy given the prevalence of a market logic in understanding the allocation of space.

A parallel argument can be made for the case of manufacturing. Changes in the economic base of cities have resulted in a growing demand for goods that can be produced by small firms that tend to benefit from a city location for access to suppliers and customers. But the general understanding about manufacturing, and one that is expressed in much policy, is that large scale and standardization are the only viable form of organizing production and hence city locations are unfeasible. Furthermore, forms of manufacturing production that do not respond to these criteria will tend to be seen as backward or marginal, conjuring up the image of the garment sweatshop. Allocating space to manufacturing activities is, consequently, viewed as inefficient. Yet the most dynamic sectors of the city's economy require access to local-based manufacturers for certain kinds of customized products.

A related question is the effect of the reorganization of income-earning activities on consumption patterns and life-style questions. There is considerable evidence showing an expansion of high-income and low-income workers. There is also evidence that a growing share of high-income jobs are going to city residents, much more so than is typical for other major cities. This new middle class has produced not only a physical upgrading of expanding portions of global cities, but also a reorganization of the consumption structure. The high income of the new workers is not sufficient to explain the transformation. Less tangible factors are considerable. An examination of this transformation reveals a dynamic whereby an economic potential—the consumption capacity represented by high disposable income—is realized through the emergence of a new vision of the good life.

The immigrants, in turn, have produced a low-cost equivalent of gentrification. Areas of the city once filled with shut-up storefronts and abandoned buildings are now thriving commercial and residential neighborhoods. The growing size and complexity of immigrant communities have generated a demand and supply for a wide range of goods, services, and workers. The separateness of the immigrant community becomes a vehicle to maximize the potential it contains. Small investments of money and direct labor in homes and shops by the individuals become neighborhood upgrading because of the residential concentration of immigrants. This upgrading does not fit the conventional notions of upgrading, notions rooted in the middle-class experience. Its shapes, colors, and sounds are novel. They, like the cosmopolitan work culture of the new professionals, are yet another form of the internationalization of global cities.

Not unrelated to these questions is the growing incorporation of women both at the high and at the low end of the occupational structure. A larger share of the new jobs created over the last decade went to women than to men; furthermore, it is estimated that over two-thirds of the job growth in the region from 1985 to 1995 will be accounted for by women. Women are 50% of all city residents employed in the business and related professional services. One question concerns the extent to which the new gentry is female. A large presence of women in the new middle-class gentrification may well lead to a renewed urbanization of families, especially female-headed or two-career families. This would engender an expanded demand for various family-type services. If we add to this the strong family base in immigrant communities, we might posit a recovery of urban space for family life that is female-led in gentrified and poverty areas and immigrant-led in some other areas of the city. Though somewhat speculative, this notion of a new, female, and immigrant-led urbanization of families is intriguing.

Finally, a question bringing these different strands together is that of the effect of economic restructuring (in its many guises) on the organization of the capital-labor relation. This is the type of question that requires theorizing alongside empirical research. It organizes much of the presentation in the next sections. Informalization and downgrading are, in the last analysis, modes of reorganizing the relationship between capital and labor. And the changes in the sphere of social reproduction described above also have contributed to such a reorganization insofar as consumption and life-style have contributed to a proliferation of small firms. These contain a distinct form of organizing work, quite different from the large-scale, standardized firm where unionization and adherence to various regulations are more typical.

BACKGROUND: SECTORAL COMPOSITION AND LOCATION OF NEW YORK CITY'S ECONOMY

While New York City is, according to many criteria, a single labor market, there are clear differences among the various boroughs and especially between Manhattan and the other four boroughs. One could

TABLE 3.1
Manhattan's Share of All New York City Jobs, 1970, 1980, 1982, 1984 (in percentages)

Industry	*1970*	*1980*	*1982*	*1984*
All industries (total)	65.8	66.0	66.7	66.1
Construction	42.8	41.2	42.9	41.5
Manufacturing	56.7	57.6	59.0	58.8
(apparel)	(67.7)	(70.3)	(69.3)	(69.24)
TCU	56.8	55.3	56.3	55.7
Wholesale	66.5	71.0	70.4	68.7
Retail	53.1	51.2	52.9	52.3
FIRE	85.9	88.4	88.9	89.7
Business services	83.0	78.1	86.4	85.7
Personal services	61.5	66.7	45.6	45.4
Health services	55.4	43.2	42.1	39.9
Educational services	69.9	64.7	64.2	63.1

SOURCE: U.S. Department of Commerce, Bureau of the Census, *County Business Patterns*, New York (selected years).

argue that, for certain industries, we are dealing with separate markets or certainly submarkets.

The distribution of economic activity by borough points to a number of trends. First, there is a disproportionate concentration (66%) of all activities in Manhattan (Table 3.1). This disproportion becomes even more accentuated when we consider certain types of activities: over 89% of finance, insurance, and real estate (FIRE) and almost 86% of business services. The extent of manufacturing concentration in Manhattan, while lower than in earlier periods, still remains high at 59%, particularly so for apparel with 69%. These figures also show that the incidence of apparel in all manufacturing is much higher than in other boroughs, where it typically is below the overall incidence of manufacturing.

A similarly uneven industry distribution is evident if we consider the metropolitan area. The location quotient for finance and business services was much higher for Manhattan than the more proximate and outer rings of the metropolitan area, though they were higher in these areas than in the city's outer boroughs (Table 3.2). Harris (1988) compared 1980 census data on place of work for the metropolitan area and Hoover and Vernon's (1962) 1956 data. He found an increase in Manhattan's location quotient for finance, a decrease in the outer

TABLE 3.2
Boroughs' Share of Economic Activity in New York City, 1984 (in percentages)

Number of Jobs in	*Total*	*The Bronx* %	*Brooklyn (Kings)* %	*Manhattan (New York)* %	*Queens* %	*Staten Island (Richmond)* %
All industries	2,953,237	5.59	12.38	66.12	14.02	1.89
Construction	93,241	9.56	16.15	41.54	29.06	3.69
Manufacturing	485,775	5.01	18.19	58.84	17.25	0.71
(apparel)	(125,568)	(3.05)	(16.04)	(69.24)	(11.07)	(0.60)
TCU	240,422	3.78	10.60	55.74	28.07	1.81
Wholesale	226,034	5.78	11.13	68.73	13.50	0.86
Retail	367,977	7.85	16.86	52.30	19.16	3.83
FIRE	514,245	1.97	3.86	89.66	3.93	0.58
Business services	264,317	2.07	4.50	85.67	6.80	0.96
Personal services	31,573	7.80	22.11	45.39	20.28	4.42
Health services	239,668	14.98	22.09	39.89	18.14	4.90
Educational services	96,073	10.80	14.60	63.10	8.98	2.52

SOURCE: U.S. Department of Commerce, Bureau of the Census, *County Business Patterns 1984*, New York.

TABLE 3.3
Industrial Specialization in the New York Metropolitan Region, 1956 and 1980

		Specialization Index (NYMR = 100)			
	Percentage Distribution	*Manhattan*	*Rest of Core*	*Inner Ring*	*Outer Ring*
1956					
manufacturing	28.2	69	121	117	128
wholesaling	6.8	145	83	68	45
finance	4.8	169	46	68	35
1980					
manufacturing	21.5	78	86	111	122
transport	9.1	102	158	81	71
wholesaling	5.1	110	94	112	78
retailing	14.1	70	101	114	113
finance	9.5	195	57	75	62
business services	5.9	144	66	95	85
personal services	2.6	108	104	200	85
professional services	22.1	87	118	97	102
public administration	4.8	104	117	88	91
NYMR	100.0	100	100	100	100

SOURCE: Harris (1988), calculated from Hoover and Vernon (1962: 248); U.S. Department of Commerce, Bureau of the Census, *Place of Work* (1984).

boroughs, and an increase in the metropolitan areas beyond the city (Table 3.3).

An examination of the sectoral and locational growth trends from 1975 to 1985 reveals the following patterns. In the mid-1970s, particularly 1975 and 1976, all boroughs lost jobs. By 1978, Queens and Staten Island were making small gains, but not Manhattan, which continued to have negative rates up to 1980. The strongest reversal in 1978 and subsequent years was in Manhattan and in Staten Island, with rates of slightly over 2% in the former and 3% to 6% in the latter. Queens went from a negative growth rate of 1.6% in 1980 to a growth of 3.1% in 1981. In 1984, for the first time in the decade, all boroughs had growth rates of 2% and over, with Manhattan weakest at 1.9%. Brooklyn went from 0.2% in 1983 to 2.8% in 1984, the Bronx from 0.3% in 1983 to 2.0%, Queens from 1.4% to 3.4%, and Staten Island from 3.3% to 4.6%.

When we disaggregate by economic sector, there is a significant change in the pattern of growth. After 1978, all boroughs had growth rates of around 4% to 6% in services. In FIRE, Manhattan had growth rates from 3.6% to 6% after 1979, a level that declined to 2.6% in 1983 and 1.5% in 1984. Queens and Staten Island also had considerable growth in FIRE after 1978, though mostly at a significantly lower level than Manhattan, until 1981, when they had a one-time growth jump of about 13% each only to return to negative rates in 1982 and 1983. Brooklyn and the Bronx continued to have negative rates until 1984 with the exception of 1981 when they had a growth rate of 1% each. In wholesale and retail trade, much of the decade saw negative rates in all boroughs, except Staten Island. There were almost no exceptions to the negative rates in manufacturing in all boroughs, the most notable being a 5% growth in Queens in 1981; toward 1984, there were small growth rates in Brooklyn and in Queens.

Aside from services, the other major growth sector in the city was construction. From 1975 to 1977, this industry experienced losses in all boroughs, with negative rates hovering around 20% in 1975 and 1976. After 1978, construction experienced significant growth in all boroughs, especially in Brooklyn, Queens, and Staten Island, and in 1980-1981 in Manhattan. As the rate of growth began to decline in Manhattan after 1982, it picked up in the Bronx and in Brooklyn.

Overall, economic activity in the city was strong. One measure is provided by the Business Activity Index, which registers the output of goods and services. This index was at a record high in 1985, with a 2.8% advance, resting largely on growth in the construction, service, and retail trade industries, and particular components of the financial industry, especially securities.

From 1984 to 1985, the city added another 51,000 jobs, following the 79,000 addition from 1983 to 1984. This is the first time since 1969 that there were two consecutive years with employment increases of over 50,000. Public and private sector employment reached 3.5 million in 1985, the highest level since 1973 (City of New York, 1986: 1).

In 1985, construction was New York City's leading growth industry, gaining 10,600 jobs over the 1984 level (compare Tables 3.2 and 3.4). Total employment in the industry stood at 105,200 jobs in 1985, the highest since 1974. Total U.S. construction employment growth was 7.3% from 1984 to 1985, compared with 11.2% for the city. When we disaggregate the information, we can see that the city's growth rate was even higher in certain branches, notably the 13.2% growth of special trade and general building contractors. The value of construction permits increased by 126% for residential permits from 1984 to 1985, by 95% for nonresidential permits, and it decreased by 8.5% for additions and alterations. The number of housing units increased by 79%. When we disaggregate the data by boroughs, some marked differences emerge. The value of residential permits increased by 283% in Manhattan compared with about 27% for the Bronx, 51% for Staten Island, and 12% for Queens. It declined by 22% for Brooklyn. The number of housing units increased by 190% in Manhattan, far above that of the rest of the boroughs. The value of nonresidential permits increased by 164% in Manhattan, almost 107% in Brooklyn, and 380% in Staten Island; the Bronx and Queens had negative rates. The value of permits for additions and alterations declined in all boroughs except in the Bronx and Queens.

The second leading growth sector in 1985 was government. Federal, state, and local employment in New York City grew by 17,800 jobs during the year, representing a 3.3% increase over the 1984 level and a significantly higher growth rate than the nation's 1.9%. This is the largest gain since 1969 and the fourth consecutive year of such growth. Over 85% of this growth was in local government, which in 1985 reached its highest level in the ten years since the financial crisis of the mid-1970s. Local government accounted for 75% of all public sector employment in New York City.

New York City continues to account for a disproportionate share of all finance jobs nationwide, with 11% of such jobs in 1985 (compared with New York City's 3.5% of the nation's work force). But growth rates in the sector have declined since the early 1980s. From 1984 to 1985, FIRE grew by only 1% in New York City, or 5200 jobs. This is the slowest increase since 1978. In contrast, nationwide FIRE employment increased by over 4%, its highest growth since 1979. As it has since 1983, banking employment continued to decline; banking accounts for a third of all FIRE jobs.

Service activities added 32,600 jobs, reaching a total of 1,038,400 jobs and accounting for almost two-thirds of the city's net employment gain in 1985. But this 3.2% increase was below the nation's 5.6%, the largest since 1978. The leading growth sectors, business and legal services, grew by over 5% each. Social, educational, recreational, and other services registered small gains while hotel employment remained the same.

For the third year in a row, the trade industry gained jobs, after severe losses throughout the 1970s and early 1980s. Having lost 90,000 jobs during that period, the city has now recovered a third of them from 1983 to 1985. Nonetheless, the 1985 gain of 1.8% is significantly lower than the national increase of over 4%. Much of the growth of the industry in the city is accounted for by retail trade, which grew at over 2%, and within retail, the largest increase was in eating and drinking establishments, which grew at over 4% and represent one-third of retail employment.

For the ninth consecutive year, the city lost manufacturing jobs. There was a loss of 15,000 jobs in nondurable and of 7000 in durable manufacturing. This brought total manufacturing jobs to 465,265, of which almost 300,000 are in the nondurable sector. All major segments of nondurable manufacturing experienced declines, unlike 1984, when textiles, printing, and publishing registered job gains. The 5.3% loss contrasts with manufacturing nationwide, which declined by less than 1% in 1985 and gained 1 million jobs during 1984.

The distribution of economic activity within boroughs shows a few pronounced differences (Tables 3.4, 3.5, 3.6). Most pronounced is the greater concentration of FIRE and business services in Manhattan, totaling 35% of all workers compared with under 10% in each of the other boroughs. A second difference is the much larger incidence of retail activity in the four outer boroughs, where it reaches about 17% and goes up to 25% for Staten Island, while in Manhattan it is under 10%. A similar pattern occurs in health services with less than 10% of Manhattan's workers compared with 11% in Queens, 14% in Brooklyn, and about 21% each in the Bronx and Staten Island. Finally, while the incidence of manufacturing in Manhattan is lower than it is in Brooklyn and Queens, that of apparel is much higher, accounting for about a third of all manufacturing compared with less than half of that for the other boroughs.

Total employment in Manhattan stood at 1,987,871. Services accounted for 35%, FIRE for 23.5%, trade for 19.4%, manufacturing for 13.7%, and construction for 1.9%. Manhattan added 35,000 jobs in 1984, after two years of little growth. Most of it was in trade and services, which accounted for 80% of job growth. Construction increased by 3.5% after two years of decline, and FIRE by only 2%. In 1985, manufacturing gained only 1%, or 10,000 new jobs.

Queens was the leading contributor to job growth after Manhattan adding 28,000 jobs to the city from 1980 to 1984. The leading growth sectors were services, trade, and construction. Manufacturing increased by 1% in 1984, but declined by 1.4% in 1985.

Brooklyn did not share as fully as other boroughs in the city's economic resurgence in the 1980s. In 1985, employment stood at 374,000. Services accounted for 35%, trade for 25%, manufacturing for

TABLE 3.4

Distribution of Economic Activities by Borough in New York City, 1985 (in percentages)[a]

	Total	*The Bronx* %	*Brooklyn (Kings)* %	*Manhattan (New York)* %	*Queens* %	*Staten Island (Richmond)* %
All industries	3,017,996	(170,483) 100	(373,640) 100	(1,987,871) 100	(426,819) 100	(59,183) 100
Construction	102,107	6.7	4.8	1.9	7.1	6.7
Manufacturing	464,265	13.7	22.6	13.7	18.7	5.8
(apparel)	(117,690)	(2.0)	(5.1)	(4.1)	(2.9)	(1.2)
Wholesale	231,938	8.0	7.0	7.9	7.6	3.3
Retail	379,338	17.6	17.3	9.9	17.0	25.8
FIRE	521,402	5.7	5.3	23.5	4.8	5.6
Business services	328,906	3.1	3.3	12.2	4.9	5.9
Legal services	58,729	0.4	0.5	2.7	0.4	0.7
Personal services	32,145	1.4	1.9	0.7	1.6	2.6
Health services	251,291	21.9	15.2	4.9	11.0	20.3
Educational services	96,734	6.2	3.8	3.1	2.0	3.6
Other services[b]	327,435	8.9	10.1	11.8	8.0	10.7

SOURCE: U.S. Department of Commerce, Bureau of the Census, *County Business Patterns 1985*, New York.

a. Percentages do not sum to 100 because other categories are not listed. Such categories include agricultural services, forestry, and fisheries; mining; and nonclassifiable establishments.

b. Other services include hotels and other lodging services; auto repair, services, and garages; miscellaneous repair services; motion pictures; amusement and recreation services; social services; museums, botanical and zoological gardens; and miscellaneous services.

TABLE 3.5

Distribution of Economic Activities by Borough in New York City, 1980 (in percentages)[a]

	Total	*The Bronx* %	*Brooklyn (Kings)* %	*Manhattan (New York)* %	*Queens* %	*Staten Island (Richmond)* %
All industries	2,902,339	(164,751) 100	(364,853) 100	(1,922,653) 100	(405,161) 100	(44,921) 100
Construction	79,852	5.38	3.41	1.72	5.75	4.84
Manufacturing	570,118	19.85	30.30	17.10	23.11	9.97
(apparel)	(148,894)	(3.37)	(6.94)	(5.45)	(3.04)	(2.17)
TCU	254,134	6.04	7.55	7.31	18.05	6.77
Wholesale	236,461	7.44	6.62	8.74	7.55	3.31
Retail	359,601	17.90	17.55	9.64	16.86	27.75
FIRE	479,345	6.75	5.73	22.05	5.02	6.77
Business services	237,425	2.98	2.83	10.63	4.10	2.69
Legal services	43,710	0.32	0.47	2.07	0.33	0.61
Personal services	30,646	1.69	1.97	0.71	1.53	2.00
Health services	201,346	16.57	12.32	4.53	8.19	19.55
Educational services	82,087	6.04	2.63	2.76	1.89	3.87
Other services[b]	304,842	8.27	7.74	11.97	6.94	10.57

SOURCE: U.S. Department of Commerce, Bureau of the Census, *County Business Patterns 1980*, New York.

a. Percentages do not sum to 100 because other categories are not listed. Such categories include agricultural services; mining; and nonclassifiable establishments.

b. Other services include hotels and other lodging services; auto repair, services, and garages; miscellaneous repair services; motion pictures; amusement and recreation services; social services; museums, botanical and zoological gardens; and miscellaneous services.

TABLE 3.6

Distribution of Economic Activities by Borough in New York City, 1970 (in percentages)[a]

	Total	*The Bronx* %	*Brooklyn (Kings)* %	*Manhattan (New York)* %	*Queens* %	*Staten Island (Richmond)* %
All industries	3,370,829	(199,304) 100	(495,605) 100	(2,219,133) 100	(423,183) 100	(33,604) 100
Construction	107,128	7.44	3.38	2.31	5.29	5.48
Manufacturing	863,796	24.72	38.62	22.11	9.67	20.65
(apparel)	(231,074)	(5.68)	(9.86)	(7.03)	(2.93)	(5.31)
TCU	354,061	14.92	9.44	9.09	17.56	6.27
Wholesale	317,092	7.20	6.56	10.56	8.15	3.93
Retail	462,295	22.12	16.93	11.06	18.72	28.55
FIRE	458,066	7.59	5.60	17.75	4.54	6.55
Business services	212,045	1.59	1.85	8.42	2.85	2.68
Legal services	45,273	.12	.25	1.24	.21	.32
Personal services	52,268	2.93	2.71	.98	2.44	2.86
Health services	150,412	5.29	6.94	3.72	4.51	11.58
Educational services	73,329	1.15	2.27	2.37	1.45	3.22
Other services[b]	275,707	4.56	5.05	10.95	4.20	7.53

SOURCE: U.S. Department of Commerce, Bureau of the Census, *County Business Patterns 1970*, New York.

a. Percentages do not sum to 100 because other categories are not listed. Such categories include agricultural services; mining; and nonclassifiable establishments.

b. Other services include hotels and other lodging services; auto repair, services, and garages; miscellaneous repair services; motion pictures; amusement and recreation services; social services; museums, botanical and zoological gardens; and miscellaneous services.

23%, FIRE for 5%, and construction for 4.8%. It did grow in 1984, adding 10,000 jobs, for a 2.8% increase. As in Queens, growth was led by the service, trade, and construction industries, which accounted for 88% of Brooklyn's job gains. Manufacturing increased slightly, the first time since 1976. Brooklyn was the only borough to have a decline in FIRE in 1984. In 1985, manufacturing declined by 6.4% (5500 jobs). All other industries added 7000 jobs in 1985, most in service, trade, and construction. FIRE increased by 1%, reversing a long-term pattern.

The Bronx did not add much to the city's job growth. In 1984, employment stood at the 1977 level of 164,000. In 1984, service accounted for 42%, trade for 25.7%, manufacturing for 13.6%, construction for 6.4%, and FIRE for 6.4%. As with Brooklyn, 1984 was a turnabout: the Bronx added 3,200 jobs (or 2%). This growth was led by the services and construction industries, which accounted for over 80% of the gain. Reversing longtime patterns of decline, there were also gains in FIRE, trade, and transportation and utilities. During 1982 and 1983, there was a severe loss of manufacturing jobs: In 1983, the Bronx lost 16% of jobs over the 1982 level; in 1984, only 2%. Job gains grew in 1985: The Bronx added 20% of the city's new jobs—over 6,000 new jobs, led mostly by construction and service industries. These two industries accounted for 88% of the Bronx's new jobs. FIRE and trade also grew; manufacturing declined by 4%, significantly less than in past years.

Staten Island had high growth in the 1980s, with strong job gains even in 1981-1982, the period of the national recession. In 1984, services accounted for 41% of all jobs, trade for 31.8%, FIRE for 6%, manufacturing for 6.1%, and construction for 6.9%. Though its total employment of 52,000 in 1984 is small, its employment growth rate of 4.6% in 1984 was the highest of all boroughs. It added 2,300 new jobs, led by FIRE and trade. There were also good gains in construction. The services industry was the weakest of all boroughs, with only a 2.4% gain. There were minor losses in manufacturing. In 1985, job gains rose by 5%, five times faster than all the other boroughs combined. Staten Island was the only borough to have job gains in all industries, even manufacturing, with 7%. Construction and trade accounted for over half of the 2,550 new jobs produced in 1985.

ELEMENTS IN THE SPATIAL ORGANIZATION OF THE ECONOMY

The industrial composition in major cities like New York is not simply a function of the general shift from a manufacturing to a service economy. Thus, while all cities contain a core of service industries, location quotients for different sized standard metropolitan statistical areas (SMSAs) clearly show the largest ones to have a disproportionate concentration of certain types of service industries, such as producer and distributive services. The locational concentration of producer services is particularly significant if we consider that the overall share of services in the GNP has grown little, rising by only 8% from 1980 to 1985 (from 62% to 70%). Indeed, except for producer, nonprofit, and distributive services, throughout the 1970s, all other service categories declined in their shares of the GNP (Ginzberg and Vojta, 1981; Singelmann, 1978; Stanback et al., 1981). In terms of employment shares, producer and nonprofit services doubled their shares of employment whereas consumer services declined. What emerges from this is that the share of services as final outputs (e.g., consumer services) is actually declining. What is increasing is the production of services as intermediate outputs—that is, producer and distributive services (Denison, 1979; Ginzberg and Vojta, 1981; Meyers, 1980; Singelmann, 1978; Stanback et al., 1981). From 1977 to 1986, the period when major shifts in the economic base of New York City were consolidated, financial services accounted for 16.7% of all the gains in the service sector and corporate services for 38.5%, according to the New York State Department of Labor. These were the leading components of service growth in the city. The next largest component in service growth was health and social services with 16.4%.

The locational concentration of producer services is in part explained by the characteristics of production of these services. These characteristics, in conjunction with the ascendance of such services in economic activity generally, both domestically and worldwide, are helpful to explain the centralization of management and servicing functions that have fed the economic boom in global cities like New York and Los Angeles. Producer services, unlike other types of services, are mostly less dependent on proximity to buyers, but benefit from proximity to other services, particularly when there is a wide array of specialized firms. Agglomeration economies occur to such firms when they locate close to others that are sellers of key inputs or are necessary for joint production of certain service offerings (Stanback and Noyelle, 1982: 17-18). Hence concentration of production in suitable locations and export, both domestically and abroad, are feasible. This would help explain why, while New York City continued to lose corporate headquarters throughout the 1970s, the number and employment of firms servicing such headquarters kept growing rapidly (Cohen, 1981; Conservation of Human Resources Project, 1977; Drennan, 1983).

By the early 1980s, New York had not only a large concentration of national service, financial, and industrial firms engaged in international business, but also a growing concentration of foreign firms. As the leading trade center for the largest single market, the United States, it emerged as a key location for foreign firms seeking access to that market. The large increase in direct foreign investment in the United States, beginning in 1981, raised the magnitude of these international transactions onto another order from what had been the case earlier. The city's vast telecommunications facilities and transportation infrastructure facilitated these developments.

Finally, New York City's role as the leading import-export center in the United States at a time of high volumes in these activities has also fed rapid growth in the wide range of associated services. At least 20% of U.S. ocean imports and 40% of air cargo imports pass through New York City. These figures represent a considerable volume of activities given that the United States accounts for 20% of worldwide imports and 15% of exports. The associated service activities range from industrial to financial.

The internationalization of production, servicing, and finance has given New York added weight as a location for key decision making about production, trade, investment, and finance by both U.S. and foreign firms. It concentrates in one particular location a strategic combination of firms, people, and resources. The importance of the U.S. market and currency feeds New York City's ranking as the leading center for international investment, where foreign and domestic investors come to place their funds both in the United States and abroad.

The level of specialization in many of the advanced services has increased. For example, large firms now tend to use several corporate, specialized legal firms. Similarly, central to the growth of management consulting has been a high level of specialization oriented to institutional investment. Thus, while cities like Boston and Los Angeles used to have management consulting sectors that compared with New York's, the changes in the financial industry have created a specific role for New York City, and one that has carried much of the industry's growth in the last few years. This high level of specialization brings about a need and a dependence on a combination of other services and resources. There is a high level of contact among firms at the point of production. These firms can serve widely dispersed regional, national, and international markets, but at the point of production, agglomeration economies are high. New York City clearly then emerges as a desirable location, notwithstanding higher costs of operation.

These developments together have brought about a rapidity and scale of growth that are specific to this economic phase, rather than being merely a continuation of older trends. This rapidity and scale have created immense pressure on space, in all its forms: housing, offices, retail, manufacturing, and space for circulation. The outcome has been a pronounced transformation in the spatial organization and built environment of the city. It has meant that the city has become a prime target for investors, both foreign and domestic.

A rather small area of Manhattan concentrates much of this activity. A distinct pattern is becoming evident with Manhattan as the key location for the financial and corporate services, and multinational headquarters, while national market headquarters are locating or moving to lower-cost locations in the greater metropolitan area. The Manhattan CBD, extending from 60th street all the way to the tip of the island, houses 60% of the city's jobs in 600 million square feet of built space, one of the highest densities among major cities. It houses more than half of all the foreign bank business in the United States. The 454 Manhattan-located headquarters listed on the Stock Exchange control US$770 billion in worldwide sales (in 1982 constant dollars). These include 38 financial and securities firms with about 100 billion in sales. In contrast, the 54 exchange-listed headquarters located in the rest of the city control $3 billion in sales. In brief, much of what constitutes the leading international center for finance, multinational headquarters and specialized servicing is situated in a rather small area in an era of global telecommunications. Because we see this pattern repeated in London and Tokyo, it may be indicative of the new forms of centralization required by the new forms of decentralization (Sassen, 1988).

One outcome of this massive increase in the volume of financial transactions and the associated services is that employment in finance and business services surpassed manufacturing in 1980. By 1984, the finance and business services employed 576,000 workers, about 140,000 more than manufacturing. As recently as 1977, there were 120,000 more jobs in manufacturing than in finance and business services. This reflects not simply a shift to a service economy, but a far more specialized recomposition in the economic base of what was once a leading light manufacturing center.

It is the possibility of concentrating production in suitable locations and exporting to other areas domestically and abroad that explains the massive development of high-rise office buildings to house these activities in major urban centers such as New York. And the large concentration of high-income jobs contributes to the explanation of the massive development of luxury residential buildings, as well as the displacement of lower-income residents. The dominant economic sectors also generate a demand, along with other sectors in the city's economy, for a range of goods and services produced by firms that lack the bidding power of firms in the economic core. These include small manufacturing firms catering to specific clients, as well as a range of what I call "industrial services": warehousing, finishing work, various kinds of delivery services, truck servicing, and packaging.

We explored some of these issues through a detailed study of Manhattan's Chelsea area, just south of midtown on the west side of the city. This area contains the largest single manufacturing-zoned district in Manhattan. The real estate industry and various government representatives have argued for years that having this zoned for manufacturing represents a waste of resources—a loss of tax revenue because there is great demand to build high-rise luxury residential and commercial buildings and little demand for the currently zoned uses. We decided to survey each building and each floor in this manufacturing zone in order to establish whether there was indeed underutilization and use by firms that did not need a Manhattan location. The study (Columbia University, 1986) compared industrial services with retail and producer services firms in the commercial area of Chelsea. For the study, we chose 13 industrial service firms, 25 retail, and 18 producer service firms. The survey focused on the following: (1) type of firm, (2) firm history, (3) employee background, (4) markets and suppliers, (5) location, and (6) terms of occupancy.

The survey revealed some distinct differences among retail and industrial services and producer services. Industrial services tend to be much older than the other two sectors, with the majority of these firms over 40 years of age. Most retail and producer services tend to be younger, with the majority of these firms established within the past 10 years. Of the producer services established within the last 10 years, many are 5 years old or less. The length of stay in Chelsea of a given firm, however, does not necessarily equal the age of the firm. This is most evident in the case of industrial services. Although only one of these firms is less than 10 years old, six of them have been in Chelsea for 10 years or less. There are different space requirements for retail, industrial services, and producer services. A large majority of retail firms used only 2500 square feet or less. At the other extreme were industrial services, most of which used 10,000 square feet or more. Most producer services used an intermediate amount of space, ranging between 2500 to 5000 square feet.

We asked firms about their relative need for a Chelsea location. Among industrial services, a majority of firms indicated that a Manhattan location was essential to the operation of their firms. Of the few industrial services that found a Chelsea location essential, this was typically because in Manhattan only Chelsea can provide them with a manufacturing space of 10,000 square feet or more.

Although most producer services indicated that a Manhattan location is essential, only a few felt the same way about a Chelsea location. In these few cases, the Chelsea location is apparently necessary because of proximity to suppliers and other firms engaged in related activities. For example, a photography studio pointed out its reliance on photo labs in the area, and a computer software design firm hires Ph.D. students from nearby New York University. We see here the formation of an agglomeration economy among the newer firms in Chelsea.

A significant number of retail firms indicated that a Chelsea location is essential, but for different reasons. For example, several mentioned that their traditional customers are in the neighborhood or simply that they had been in the area a long time. Even a national fast-food chain franchise felt that it was dependent on a Chelsea location. This suggests a neighborhood effect among the retailers. We also obtained information on prior uses of the space in which retail, industrial services and producer services are now located. Most producer services are located in space previously used for printing, garment activities, or other manufacturing uses. Industrial services are often located in space that was formerly used for industrial or manufacturing purposes, such as a warehouse, refrigerated storage, or weaving and knitting. In general, retail firms replace retail firms, industrial services replace industrial services, but producer services are replacing manufacturing uses.

Analysis of the markets served by these firms reveals further differences among the three sectors. As expected, a large majority of retail firms have a market that extends no further than the immediate neighborhood of Chelsea. Many of the industrial services have a market that extends throughout the five boroughs of New York City and into the tristate area of New York, New Jersey, and Connecticut. Producer services tend to serve Manhattan or else extend far beyond that to include places all over the United States and overseas. It should be noted that the market areas are not always progressively inclusive. A Chelsea-based firm with a national market does not necessarily have clients in every market delineated below that level (i.e., tristate, New York City, Manhattan, and Chelsea). One public relations firm included in the survey had clients in California, but not a single one in New York City. It was located in Manhattan in order to have easy access to advertising firms on Madison Avenue.

The range of land uses that the new economic core requires carries with it spatially expressed contradictions and hence potential for conflict. On the one hand, there is an expanded effective demand for highly priced commercial and residential space. On the other, there is a demand for lowly priced manufacturing, commercial, and residential space. If those who demand space but have less bidding power are either (1) not part of the new economic core or (2) part of it but can locate in more distant areas where prices are lower, then we are merely confronting a process of displacement. If a critical mass, or a significant share of those with lower bidding power, however, are part of the new economic core and/or need proximate locations to the more central components of that core, then we may posit that this is a process whose territorial expression contains a contradiction that may affect the functioning of the economic core. This would be reminiscent of some of the outcomes evident in the case of Silicon Valley, though, given New York City's complexity, it would be impossible to attain the degree of analytical clarity feasible in the former. In this context, a manufacturing zone such as Chelsea's functions as a subsidy for firms with lower bidding power that need a Manhattan location and that supply a needed good or service to core firms.

The concrete components of the economic base that engender this range of land uses also include trends in the sphere of social reproduction. The major economic sectors in the city create an expansion in the supply of high- and low-income jobs. This translates into parallel discontinuities in the demand for housing, types of food, clothing, furniture, restaurants, and other components of household and personal consumption. The city needs to accommodate both expensive and lowly priced entities meeting these two types of demand. In certain cases, furthermore, these have to be provided in the same location. Food places need to be available for both the highly and the lowly paid workers at lunchtime, and transportation has to bring both types of workers to the same building or same general location. If a city's work force is dominated by a large middle-class income stratum, this range of needs can be more easily accommodated because there will be a tendency for a middle-priced operation to prevail or workers will have more leverage to get the somewhat less profitable needs accommodated.

But the type of disparity in consumption capacity, and probably political influence, we see emerging today in New York City will tend to lead to a skewed response by the economy and the public sector. The large high-income work force will have a disproportionate effect on both and thereby maximize the squeezing of the lower-income work force and the restaurants, shops, and transportation vehicles that service them. In the case of the housing market, there has been direct displacement of lower-income households leading in the extreme to homelessness. In general, the outcome has been a highly segmented use of urban space: high-income gentrification of a rather sizable area of the city (compared with what was the case in the early and middle 1970s), increased poverty and public sector neglect in lower-income areas, and a growing concentration of immigrant-occupied areas. The solid working- and lower-middle-class areas of the city found in the outer boroughs have tended to be stable in certain cases, to suffer encroachment of higher-income households in others, or to experience disintegration as residents reach retirement or death and there is no cohort of similarly placed workers to replace them.

JOB SUPPLY

The consolidation of this economic core of top-level management and servicing activities needs to be viewed alongside the general move to a service economy and the decline of manufacturing. New economic sectors are reshaping the job supply. So are new ways of organizing work in both new and old sectors of the economy, however.[1] Components of the work process that even 20 years ago took place on the shop floor and were classified as production jobs today have been replaced by a combination of machine/service or worker/engineer. Activities that were once all consolidated in a single-service retail establishment have now been divided between a service delivery outlet and central headquarters. Finally, a large array of activities that were being carried out under standardized forms of organizing work a decade ago are today increasingly characterized by informalization, such as sweatshops and industrial homework. In brief, the changes in the job supply evident in major cities like New York are a function both of new sectors *and* of the reorganization of work in both the new and the old sectors.

At the national level, the general trends shaping the job supply have brought about a greater inequality in the earnings distribution of workers over the last decade. The shift to a service economy is, first of all, generally recognized to result in a greater share of low-wage jobs than is the case with an economy dominated by a strong manufacturing sector (Singelmann, 1978; Stanback et al., 1981; Bluestone et al., 1984; Jacobson, 1978). Second, some of the fastest-growing service industries are characterized by larger than average concentrations of lowly and highly paid jobs (Stanback et al., 1981; Appelbaum, 1984). Third, there has been what I call a downgrading of manufacturing jobs: major new industries, notably electronics, have a high incidence of low-wage jobs in production and assembly, while several of the older industries have undergone a social reorganization of the work process characterized by growth in nonunion plants and a rapid increase in sweatshops and industrial homework (Balmori, 1983; Marshall, 1983; New York State Department of Labor, 1982a, 1982b; Sassen-Koob, 1981; Wolff, 1984). Fourth, the technological transformation of the work process, in part underlying the above trends, has further added to earnings polarization by either upgrading or downgrading a vast array of middle-income jobs. Mechanization and computerization have transferred skills to machines and have shifted certain operations from the shop floor to the computer room or designer's studio.

Various kinds of evidence document aspects of this growing polarization. Using 1970 and 1980 census data on earnings organized by industry-occupational cells, I found that the two highest earnings classes increased their total share from 32.2% to 37%, while the two lowest ones increased their share from 32% to 38.5% (see Table 3.7). Correspondingly, the two middle classes reduced their share by 11 percentage points. Estimates made by the Bureau of Labor Statistics (U.S. Department of Labor) show that the highest absolute growth in jobs will occur in low-paying occupations. Using unpublished data and forecasts from the Bureau of Labor Statistics, Bluestone et al. (1984) examined the 1969-1982 growth performance of 136 manufacturing and non-manufacturing industries monitored regularly by the bureau and categorized them by their mean annual wage in 1980. They found that, from 1969 to 1982, the industries with the largest net growth in jobs were those that in 1980 paid the lowest average wages. While 63% of net new jobs were found in industries with a 1980 average wage of less than $12,500, there was almost no net growth in industries with an average 1980 wage of $22,000 or more.[2] Projecting these trends into the 1980-1995 period suggests an even more pronounced concentration of net job growth in low-wage industries. Finally, the research by Jacobson (1978) on what happens to workers who are displaced from basic manufacturing industries shows that a majority may face a permanent loss in earnings. Bluestone and Harrison (1982: 97) also found a loss in earnings among New England workers after displacement from tradi-

TABLE 3.7
Distribution of U.S. Labor Force Among Earnings Classes, 1970 and 1980[a]

Earnings Classes[b]	1970 Total	1970 Total (pairs combined)	1970 Female	1970 Male	1980 Total	1980 Total (pairs combined)	1980 Female	1980 Male
1.60 and above	11.3	32.2	7.5	9.4	12.9	37.0	4.8	11.0
1.59 to 1.30	20.9		18.6	18.9	24.2		14.5	20.7
1.29 to 1.00	18.9	35.8	21.5	23.1	12.8	24.5	12.8	15.6
.99 to .70	16.9		10.5	14.3	11.7		15.8	17.0
.69 to .40	22.8	32.0	13.5	15.4	25.2	38.5	16.7	11.8
.39 and below	9.2		28.4	19.0	13.3		35.4	23.9

Distribution of Total U.S. Labor Force (in percentages)

SOURCE: Based on U.S. Department of Commerce, Bureau of the Census (1972, 1982).
a. Civilian workers 14 years and over by total money earnings.
b. Earnings classes are derived from the application of 1975 average earnings for each major occupation within each industry group. A basic assumption is that the relative income at 1975 levels for each occupational industrial subgroup is constant—in this case, from 1970 to 1980. I followed the method used by Stanback and Noyelle (1982: chap. 3) in their comparison of 1960 and 1975 earnings for industry-occupation cells. The total earnings distribution obtained is then divided into sextiles. The major industry groups are manufacturing, construction, distributive services, retail, producer services, consumer services, nonprofit services (health and education), public administration. Not included are agriculture, fisheries, and mining. The major occupational groups are professional, technical, manager, office clerical, nonoffice clerical, sales, craft workers, operatives, service workers, and laborers.

tional mill-based industries; a shift to the service sector, where job growth is occurring, entailed a decline in earnings for most of these workers.

All these trends are operating in New York City, in many cases with greater intensity than national average data describe. This greater intensity can be rooted in at least three conditions. First, there is the already mentioned locational concentration of major growth sectors with highly polarized income distributions. Second, there is a proliferation of small, low-cost, service operations made possible by the massive concentration of people in such cities, in addition to a large daily inflow of nonresident workers and tourists. The ratio between the number of these service operations and the resident population is most probably significantly higher in New York than in an average city.[3] Furthermore, the large concentration of people in major cities tends to create intense inducements to open up such operations as well as intense competition and very marginal returns. Under such conditions, the cost of labor is crucial and hence the likelihood of a high concentration of low-wage jobs. Third, for these same reasons, together with other components of demand, the relative size of the downgraded manufacturing sector would tend to be larger in a city like New York than in average-sized cities. Indeed, in many of these cities, such a downgraded manufacturing sector is not a significant factor, if present at all (Sassen, 1988).

When we consider the evidence on different occupational and earnings distributions for various industries in combination with the locational patterns of such industries, it becomes rather clear that major cities like New York and Los Angeles have a high incidence of industries with considerable income and occupational polarization.

The core economic sector in New York City consists of a wide array of producer services, from banking to management consulting and business services, which accounted for 31% of all employment by 1985 (see Tables 3.2 and 3.8). The high incidence of producer and other services in New York City's employment structure contributes to the generation of low-wage jobs directly and indirectly. The data on earnings classes show a very high incidence of the next to lowest earnings class in all services, except distributive services and public administration (Stanback et al., 1981). Only 17% of workers in manufacturing and 18.8% in construction were in the next to lowest earnings class, compared with almost half of all workers in the producer services (Stanback and Noyelle, 1982: 33). The other half of the workers in producer services were in the two highest earnings classes. Only 2.8% of the workers in these services were in the middle earnings class, compared with half of all workers in construction and manufacturing.

County Business Patterns data on weekly payments by industry in New York City reveal two trends (see Table 3.9). First, there is considerable variation in such payments among the major industry groups. Second, Manhattan-based jobs have higher average weekly payments in all major industry groups than jobs in the outer boroughs of the city. These data are inadequate in that they are averages and do not indicate the distribution of weekly payments by occupation within industry. In 1985, nominal weekly payments in construction ranged from a high of $689 in Manhattan to a low of $468 in Brooklyn (Kings County); in manufacturing, from $575 in Manhattan to $342 in Brooklyn; in FIRE, from $732 in Manhattan to $344 in the Bronx; in services as a whole, from $487 in Manhattan to $314 in Staten Island (Richmond County). Within the services sector, business services ranged from $501 in Manhattan to a low of $242 in Staten Island;

TABLE 3.8
Employment in Producer Services in New York City, 1970, 1977, 1982, 1985-

	Employees			
SIC Industry	*1970*	*1977*	*1982*	*1985*
60 Banking	129,267	137,253	170,201	158,400
61 Credit agencies	14,622	18,537	21,719	27,674
62 Commodity brokers	95,198	68,713	93,466	118,896
63 Insurance carriers	87,799	73,452	71,436	66,849
64 Insurance agents	30,073	23,134	26,678	30,889
65 Real estate	100,127	87,424	87,754	94,022
66 Combined real estate and insurance[a]	1,334	1,117	665	613
67 Holding, investment office	8,297	19,657	16,878	17,963
73 Business services	212,045	193,043	250,317	283,906
81 Legal services	29,997	37,004	48,962	58,729
86 Membership organizations	81,885	54,220	52,750	56,754
89 Miscellaneous business services	59,065	47,314	67,081	65,585
Total	839,709	760,868	907,907	980,280
Total for all New York industries	3,370,829	2,714,385	2,896,979	3,017,996
Share of producer services in all industries	25.0%	28.0%	31.3%	32.5%
Employment growth rates		1970-1977	1977-1982	1982-1985
		−9.4%	19.3%	8.0%

SOURCE: U.S. Department of Commerce, Bureau of the Census, *County Business Patterns*, New York (selected years).
a. Figures are from an average range of employees (e.g., 100-249).

personal services were uniformly low, ranging from $257 in Manhattan to $176 in Staten Island; and legal services ranged from $729 to a low of $405 in Queens.

To obtain a measure of the *direct* creation of low-wage jobs in the major service industries, I used data from the occupational survey done by the New York State Department of Labor (1979, 1980) and identified the full array of a particular *subtype* of low-wage jobs (Sassen-Koob, 1984a: see table 2). These are low-wage jobs with few if any skill and language requirements and little unionization—that is, jobs that

TABLE 3.9
New York City: Average Weekly Payments by Industry and Borough, 1985 (in US $)

1985	*Bronx*	*Brooklyn (Kings)*	*Manhattan (New York)*	*Queens*	*Staten Island (Richmond)*
Construction	554	468	689	597	493
Manufacturing	400	342	575	405	507
Apparel	190	227	352	266	197
TUC	589	478	698	653	554
Wholesale	468	413	656	492	416
Retail trade	240	245	322	242	213
FIRE	344	356	732	371	369
Services	358	316	487	325	314
business services	286	272	501	295	242
personal services	223	244	257	210	176
legal services	514	475	729	405	462

SOURCE: U.S. Department of Commerce, Bureau of the Census, *County Business Patterns, 1985.*

promote disenfranchisement among workers and could conceivably be held by foreign language, powerless workers. Overall, 16% of all jobs in the major service industries covered by the survey fit into this category of low-wage jobs. Almost 11% of all jobs in FIRE were of this kind, 24% of all jobs in business services, and 18% in the remaining service industries could be categorized in this way. Also worth noting is that the highest incidence of such jobs is found in the fastest-growing employment sector of the city, business services.

There is also an *indirect* creation of low-wage jobs induced by the presence of a highly dynamic sector with a polarized income distribution. It takes place in the sphere of consumption (or social reproduction). The expansion of the high-income work force, in conjunction with the emergence of new cultural forms in everyday living, has led to a process of high-income gentrification that rests, in the last analysis, on the availability of a vast supply of low-wage workers. As I have argued at greater length elsewhere (Sassen-Koob, 1981), high-income gentrification is labor intensive, in contrast to the typical middle-class suburb, which represents a capital-intensive process—tract housing, road and highway construction, dependence on private automobile or commuter trains, heavy reliance on appliances and household equipment of all sorts, and large shopping malls with self-service operations. High-income gentrification directly and indirectly replaces much of this capital intensity with workers. Behind the gourmet food stores and specialty boutiques lies an organization of the work process that differs from that of the self-service supermarket and department store. Similarly, high-income residences in the city depend to a much larger extent on hired maintenance staff than the middle-class suburban home, with its heavy input of family labor and of machinery from household appliances to lawn mowers.

The differences in the organization of the work process are evident in both the retail and the production phase (Sassen, 1988; Morales and Mines, 1985). High-income gentrification generates a greater demand for goods and services that are customized and less likely to be mass produced or sold through mass outlets than middle-class suburbanization. Customized production and limited runs of production tend to be associated with labor-intensive methods of production and sold through small, full-service outlets. Subcontracting part of this production to low-cost operations, be they sweatshops or households, is common. The overall outcome is an increase in low-wage jobs and in small firms for production and retailing. The large department stores and supermarkets, on the other hand, need large quantities and standardized products; they tend to buy from mass producers, often located in other cities and states, with a national market. Mass production, standardization, and mass distribution outlets facilitate unionization both in production and in sales work.[4] Customized production, limited runs, and small retail outlets promote informalization.

The magnitude of the expansion of high-income workers and the high levels of spending contribute to this outcome. New York City, like all major cities, has long had a core of wealthy residents or commuters. While this core has probably been expanded by a large influx of wealthy foreigners, by itself, it could not have created the large-scale residential and commercial gentrification in the city. As a stratum, the new high-income workers are to be distinguished from this core of wealth, or upper class. Their disposable income is, mostly, not enough to make them into major investors. It is, however, sufficient for a significant expansion in the demand for highly priced goods and services; that is to say, to create a sufficiently large demand so as to ensure economic viability for the producers and providers of such goods and services. Furthermore, the level of disposable income is also a function of lifestyle and demographic patterns, such as postponing having children and the larger number of two-earner households.

THE DOWNGRADED MANUFACTURING SECTOR

A second major source of low-wage jobs is the expansion of a downgraded manufacturing sectors. The social and technical transformation of the work process, as well as the new production jobs being created by the electronics industry have all contributed to an increase in low-wage blue-collar jobs in manufacturing (Appelbaum, 1984; Bluestone et al., 1984). At the same time, industries with well-paying blue-collar jobs are losing many jobs and downgrading others. Familiar instances of this downgrading are the proliferation of sweatshops and industrial homework, not only in garments (Abeles, Schwartz, Hackel & Silverblatt, Inc., 1983; Waldinger, 1985), but also in footwear, furs, furniture, and electronics (New York State Department of Labor, 1982a, 1982b). Also indicative is that one of the fastest-growing industries, electronics, has characteristics of production that allow for the shifting of production and assembly jobs to low-wage countries, where it employs mostly unskilled workers (Nash and Fernandez-Kelly, 1983; Safa, 1981). The same characteristics of production shape the job supply in electronics production and assembly in the United States. Of interest here is the sharp drop in levels of unionization over the last decade in areas with rapid growth in high-technology industries, such as Los Angeles and Orange Counties (Wolff, 1984).

The consolidation and expansion of a downgraded manufacturing sector in New York City is the result of several concrete developments aside from the more general processes of social and technical transformation cited above.[5] First, labor-intensive industries were affected differentially by capital flight in the city. In the case of the garment industry, the largest employer in the city's manufacturing sector, the bigger shops with mechanized production were the ones that moved (NACLA, 1978; Waldinger, 1985). The less-mechanized branches, the specialized shops, and the industry's marketing and design operations have remained in the city. Furthermore, the changing structure of consumption may also affect the city's garment industry: The greater demand for specialty items and limited edition garments promotes the expansion of small shops and industrial homework in the city, because small runs and proximity to design centers are important locational constraints (Sassen-Koob and Benamou, 1985). A parallel argument can be made for other industries, notably furniture, furs, and footwear.[6]

Second, there has been a considerable influx of direct foreign investment in manufacturing in the New York-New Jersey metropolitan region over the last decade (Bauer, 1980; New York Chamber of Commerce and Industry, 1983; Port Authority of New York and New Jersey, 1981, 1982). The crucial postcrisis period from 1976 to 1980 shows New York State to have experienced the largest net increase in foreign-owned plants (both acquired and newly built) of all states; 196 compared to 161 in California, 94 in North Carolina, and 87 in Texas. The figure for the New York-New Jersey metropolitan area was 88. Nationwide, New York and New Jersey had 18.6% of all foreign-owned manufacturing plants in 1980. These figures are an undercount because they only include plants with 55 or more employees. The New York-New Jersey region is known to have a large number of smaller plants. While there are many factors promoting the desirability of investment in the area, one that is of significance to this analysis is the availability of a suitable labor supply: a skilled industrial work force that is politically weakened by capital flight and a low-wage, politically powerless immigrant work force.[7] At the limit, the availability of the latter means that investment in this region (or the Los Angeles region, for example) becomes an alternative to investment in offshore locations, particularly if the final market is the United States (Sassen, 1988).

Third, small-scale, immigrant-owned plants have grown rapidly in number due to easy access to cheap labor and, most important, a growing demand for their products in the immigrant communities and in the city at large. Probably the clearest illustration of this expansion is Chinatown (Abeles, Schwartz, Hackel & Silverblatt, Inc., 1983; Wang, 1979). Capital from the overseas Chinese has been flowing into New York City's Chinatown at a particularly accelerated pace since the middle of the 1970s. Aside from real estate, the key growth factors in Chinatown's boom are the garment and restaurant industries. While overall recorded employment in New York City's garment industry has been declining throughout the 1970s, it actually increased in Chinatown. The number of unionized jobs even grew, doubling from 1970 to 1982 (Abeles, Schwartz, Hackel & Silverblatt, Inc., 1983: 3-5). Activity has been so intense that a growing number of shops owned by Chinese are spreading into the neighboring boroughs of Queens and Brooklyn. Immigrants from many different countries have set up small operations in food processing and furniture, manufacturing a variety of consumer items and, most important, garments. The needed capital is typically obtained through pooling of family and friends' resources or an arrangement reminiscent of "rotating credit associations."

The expansion of a downgraded manufacturing sector involves, to a considerable extent, the same industries that used to have highly organized plants and reasonably paid jobs, but it replaces these with different forms of production and organization of the work process, such as piecework and industrial homework. It also involves new kinds of activities, however, that are associated with the new major growth trends. There seem to be three kinds of activities in this regard, and they are not always mutually exclusive. First, there is some evidence that an increasing component of manufacturing has a specific, narrow clientele: It ranges from the production of tools and goes through furniture to boxing gloves for particular firms (Conference Board, 1981; Port Authority of New York and New Jersey, 1981). Second, there is the small-scale production for retailers selling to the growing high-income population stratum—whose numbers have increased sufficiently to promote the creation of a whole array of small businesses catering to their tastes. What makes this production profitable is, undoubtedly, the possibility of gaining access to cheap labor and having huge markups on final prices. Subcontracting to small plants willing to manufacture a limited number of items is a key strategy to ensure the viability of this type of limited market production. Third, production for the newly expanded low-income mass market, with much of the merchandise sold on the street, has partly replaced production for the middle-class market. The locational patterns of plants producing for middle-income markets are probably somewhat different, because there is access to a multiplicity of outlets that sell for a nationwide market such as department and chain stores. This facilitates production through large plants. The low-income mass market that is emerging in large cities like New York and Los Angeles is, to a considerable extent, supplied by sweatshop production located in these cities themselves or overseas. The existence of a large immigrant work force facilitates sweatshop production close to the final market, something that is particularly desirable in the case of bulkier items (such as pillows) and perishables. It is not possible to obtain measures of these three kinds of activities; this is merely a tentative interpretation of the fragmentary evidence available (see Sassen-Koob, 1981).

The possibility for manufacturers to develop alternatives to the organized shop and to large-scale standardized production becomes particularly significant in growth sectors. The consolidation of a downgraded manufacturing sector, whether through social or technical transformation, can be seen as a politico-economic response to a need for expanded production in a situation of growing average wages and militancy, as was the case in the 1960s (see Nash and Fernandez-Kelly, 1983: 12-23). While it is true that average wages in production in New

York City never were among the highest in the country, given the absence of such key industries as steel, auto, or aerospace, it is also true that average wages were increasing and reached one of their higher relative levels as recently as 1970, when they were 101.2% of the national average hourly wage in production. One indicator of the extent of downgrading is the decline of this wage level, which fell to 87.6% of the national average in 1982. It would be significantly lower if sweatshops and industrial homework were included. This decline can also be seen in an area like Los Angeles, whose economic base is radically different from that of New York City, because it is dominated by aerospace and electronics. Hourly production wages in Los Angeles fell from 108.8% of the national average in 1970 to 100.7% in 1982. Feeding into the decline in wage levels in Los Angeles are both new manufacturing sectors, notably electronics production and assembly, and old sectors such as the proliferation of sweatshops in the garment industry (see Sassen-Koob, 1984a).

A good part of what I am here calling the downgraded manufacturing sector is an instance of informalization, or a component of the informal sector.[8] We need to distinguish two spheres for the circulation of goods and services produced in the informal sector. One sphere circulates internally and meets mostly the demands of its members—that is, small, immigrant-owned shops in the immigrant community that service the latter; the other circulates throughout the "formal" sector of the economy. In this second case, informalization represents a direct profit-maximizing strategy, one that can operate through subcontracting, the use of sweatshops and homework, or direct acquisition of goods or services. This second aspect is somewhat lost when we use the concept of the informal sector, while it comes to the fore with the notion of a downgraded manufacturing sector—and, increasingly, downgraded mass consumer services, whether public or private, alongside increasingly upgraded non-mass consumer services.

INFORMALIZATION TRENDS IN NEW YORK

A small but growing body of evidence points to the expansion of informal sectors in major cities of the United States over the last decade (Columbia University, 1987; Fernandez-Kelly, forthcoming; Sassen-Koob, 1981, 1984b; Sassen-Koob and Grover, 1986; Stepick, 1984).[9] These studies are mostly based on ethnographic research. They are to be distinguished from studies that aim at overall estimates of the underground economy based on aggregate figures for the supply and circulation of money (Gutmann, 1979; Spitznas, 1981; Tanzi, 1982). As categories for analysis, the underground economy and the informal sector overlap only partly. Studies on the underground have sought to measure all income not registered in official figures, including income derived from illicit activities such as drug dealing. Studies on the informal sector focus on the production and sale of goods and services that are licit, produced, and/or sold outside the regulatory apparatus covering zoning, tax, health and safety, minimum wage laws, and other types of standards.

My current research on informalization in New York City attempts to specify a relationship between (1) the growth of informal activities in New York City and (2) overall transformations in the economy and the existing regulatory environment. The research will serve to illustrate the dynamics of informalization. Individual motivations and diverse structural forces contribute to growth in New York's informal economy, among them business' and employees' desire to evade taxes and regulatory costs, and the growth in the immigrant population. A central question for research, however, is whether transformations in the organization of production and consumption induce informalization in that they create business and employment opportunities that are highly susceptible to informalization.[10]

This emergent or developing opportunity structure may or may not be facilitated by the regulatory system. Modes of government regulation are particularly important for understanding the rise of informal production because of the costs that they impose on formal businesses through their various licensing fees, taxes, and restrictions. Labor costs also have an effect on the formation and expansion of the informal sector: directly, in terms of the wages paid, and indirectly, in terms of various contributions by law. These inducements to informalization vary according to industry and location.

The objectives of the larger and ongoing research project are the following: (1) to assess the extent to which informal economic activities are growing in New York City and in which kinds of communities and industrial sectors; (2) to identify the linkages between these informal activities and formal sector activities (in the communities, in the industries, and in the wider city economy); (3) to identify whether there exists, and, if so, compare the weight of demand that is external to the communities in which informal goods are produced versus that which is internal to the communities; (4) to identify the role that informal production plays in the spatial organization of particular industries in

New York City; (5) to examine the importance of New York City as a location to informal producers; and (6) to identify existing government policy toward, and those which implicitly affect, informal producers.

The research involves a combination of (1) secondary data analysis, (2) ethnographic research in select communities and various kinds of workplaces, and (3) interviews with informed individuals, including local planning officials, union officials, and community members. We used data on occupational safety and health administration violators, along with overtime minimum wage legislation violators to identify trends in informal production in New York City. On the basis of these data, we targeted certain industries for more in-depth study of the informal component of their total production chain. We did fieldwork in communities identified by knowledgeable people as having a large informal sector or having a high density of immigrant population as an indicator of good probability of informal activity. We used interviews with community boards and local development corporation representatives to target specific sites with violations. (Complaints of businesses in violation of legally defined uses of building or zoning ordinances are normally registered with local development corporations and community boards before being referred to the Department of Buildings.) Zoning maps and data from the Department of Buildings were used to obtain more detailed information on informal producers. Our site visits and interviews helped us identify the extent and kinds of informal activities in these communities. They also helped us to assess how existing regulations influence informal producers and to determine if new modes of regulation are needed.

On the basis of our secondary data analysis, fieldwork, and interviews, we found the following profile of the informal sector in the New York City area (Sassen-Koob and Grover, 1986): (1) a rapid proliferation of informal operations (including industrial homework); (2) a rather wide range of industrial sectors with such operations—apparel, general construction contractors, special trade contractors, footwear, toys and sporting goods, electronic components, and accessories; (3) to a lesser degree, operations in particular kinds of activities like packaging notions, lamp shades, artificial flowers, jewelry, distribution activities, photo engraving, and manufacturing of explosives; (4) a strong tendency for such operations to be located in densely populated areas with very high shares of immigrants; and (5) an emergent tendency for "traditional" sweatshop activity (notably garments) to be displaced from certain areas undergoing partial residential and commercial gentrification and in its stead new forms of unregistered work catering to the new clientele.

These preliminary findings are based upon extensive interviews and intensive fieldwork in six industries (Sassen-Koob and Grover, 1986): construction, garments, footwear, furniture, retailing, and electronics. In construction, it has been estimated that 90% of all interior work in New York City is done without a building permit.[11] Although there is little underground activity in large public works projects, there is much illegal work in rehabilitation and small-scale store, apartment, and commercial renovation. One informant noted that most highly specialized craftsmen are immigrants, notably cutters, masons, and plasterers; many do not have residence or work permits.[12] In addition, there is a widespread use of subcontractors for jobs that were once done through direct hiring of union labor; many subcontractors are not fully registered, and a growing number of them are Hispanics in an industry where they were once excluded.[13] Subcontracting is one way for large companies and large construction projects to reduce their costs and avoid taxes on some earnings.[14] Finally, unregistered work is common in renovations and alterations work in offices.[15] Second, several studies document the rapid increase in the number of sweatshops (both union and nonunion) and industrial homework in the garment industry since the early 1970s.[16] Overall, the following can be noted for the New York area: (1) there are an estimated 3000 garment sweatshops; (2) they have been spreading to peripheral areas within the city and outside the city as rents have increased—for example, out of the Chinatown area in Manhattan and into northern Manhattan, Brooklyn, and Queens; (3) sweatshops and homework are most common in women's and children's clothes; and (4) new branches in which sweatshops and homework are increasing rapidly are the so-called "knitting mills," furs, embroidery, stuffed toys, and clothing for toys. Third, in the footwear industry, several informed people from city agencies and unions said that unregistered work is spreading both in rather standardized production and in highly crafted production (sandals, moccasins). One estimate is that 10% of all shops are unregistered; and unregistered work is also increasing among union shops, via subcontracting and homework, a recent development in the last five to ten years. Many "fly-by-night" operations stay open only for six to nine months.[17] In addition, there is also growing evidence of unregistered work in handbags and other leather goods.

Fourth, in the furniture industry, both large, standardized factories and certain branches of furnishings (bedding, mattresses) have moved out of the area, primarily to the South. The new manufacturing activity in furniture and craft work is commonly located in areas not zoned for this work. Examples are Ridgewood, Astoria, in Queens, and Williamsburg, Sunset Park, in Brooklyn.[18] There are several new manufacturing shops, employing immigrants almost exclusively, making high-cost, crafted cabinets and other furniture in Jackson Heights (Queens) geared to the local increasingly middle-income residential population and fancy retail shops associated with commercial and residential gentrification.[19] In fact, as an area gentrifies, there is a very localized demand for on-the-spot woodwork and installation.

Fifth, in retailing, there is growing piecemeal evidence of unregistered retail activity in more and more branches of industry, such as in jewelry and furs, flea markets, and among peddlers.[20] Several of the other sources already mentioned cited evidence of informal buying and selling of inputs and outputs. Last, in electronics, there is growing evidence that unregistered work is increasing.[21] Despite the need for more fieldwork, certain conclusions are emerging: (1) homework is also being done by native, Anglo workers; (2) subcontracting is central to the electronics industry in the New York-New Jersey area among small firms with highly specialized kinds of production and specific customers;[22] (3) there are chains of subcontractors, and often, at some point along the way, there is a "garage-shop" or "basement-shop" or "neighbor" who is part of the subcontracting and often unregistered.[23]

SPATIAL ORGANIZATION

In addition, several distinct patterns are emerging in the spatial organization of the informal sector. For certain industries, the informal sector is basically located in immigrant communities. Furniture making, on the other hand, may well be located close to the final market in gentrifying areas. A third pattern is represented by internal renovation, where the workplace changes with every contract. Yet another kind of indicator emerged from the transportation industry. In the case of gypsy cabs, or liveries, their "workplace" tends to be confined to areas of the city where the regular cabs refuse to go. In the case of vans, we found two patterns, both clearly linked to the trip to work. One was the established public transportation routes; the other linked immigrant communities in the outer boroughs to areas with a high concentration of low-wage service and retail jobs in Manhattan. Four of the five New York City boroughs—Manhattan, Brooklyn, the Bronx, and Queens—were found to have areas with significant concentrations of informal activities. Furthermore, we found a tendency toward spatial concentration of certain types of activities in particular locations. For example, auto repair shops were concentrated in Sunset Park in Brooklyn, in Corona-Elmhurst in Queens, and in Washington Heights in Manhattan. The pattern was akin to an industrial district. Such areas then emerge as major servicing areas for local and nonlocal users.

Another, more complex, pattern involves a tendency toward spatial concentration of a wide array of manufacturing and related servicing activities in several key locations. For example, in West Astoria (Queens), we found shops doing glass work for buildings and vehicles, refinishing of restaurant equipment, auto repair, garment shops, and carpentry shops that make the framework for furniture that is then sent to other locations for finishing (probably both formally and informally), and cabinet-making shops. All of these various shops were operating informally. This area has emerged as a significant manufacturing zone. City government has opted to disregard the violations because of the decline of space available for manufacturing in Manhattan. The significant concentration of manufacturing shops has brought about a whole array of related service shops, as well as creating a zone where new industrial uses are occurring. For example, a major tile retailer in the city illegally warehouses the tiles in West Astoria. One major implication is that what may initially be a small number of manufacturing shops operating informally, may, under certain conditions, develop into an industrial district with agglomeration economies that will draw an increasing number of industrial users. This then becomes a de facto manufacturing zone.

Yet another pattern is the formation and expansion of a new industrial district in an area that was once a thriving industrial zone but has gone into severe decline. This is well illustrated by the case of Williamsburg in Brooklyn. This area is near to what was once a thriving navy yard. Much of the industry associated with the original industrial uses went out of business or left the area. The largely vacant industrial spaces eventually were taken over by different industrial uses than those that had originally occupied the buildings. Now we see there a growing concentration of informal manufacturing in garments, handbags, leather bags, knitting mills, and confectionery. We also noted seasonal-type subcontracting by brand name cosmetics firms to package Christmas and other special orders. The work involves mainly assembly-type operations.

Finally, in some areas undergoing rapid commercial and residential gentrification, we found manufacturing shops catering to the needs engendered by that process. Jackson Heights in Queens illustrates this well. The number of stores has grown rapidly and, given high rents, need to be highly profitable. Much of the retail market caters to the new population of high-income residents, and this process has also created a need for various "inputs" whose production would be unlikely to generate the high incomes needed to pay the high rents of street-level spaces. One of these key "inputs" is cabinet and related wood installations. We found numerous shops in the area producing mostly unfinished, but very high quality, customized furniture and kindred products. These shops were above storefronts rather than on street level and, most important, they were in residential spaces. They employed immigrant workers, mostly highly skilled craftsmen.

One of the trends that stood out in our community studies is that several of these areas with manufacturing activity or industrial services were zoned for residential and/or commercial uses, and not for manufacturing. This was the case for the West Astoria location described above, for Jackson Heights, and for the auto repair shops in Sunset Park. To some extent, these areas are also largely occupied by immigrant residents, many of whom are employed in these informal shops. Most generally, what we are seeing is a pattern of land use that does not necessarily correspond to zoning regulations. Regulations, in this case zoning laws, are one variable that needs to be introduced alongside others into the analysis of informalization. When the magnitude of the concentration of informal shops reaches that of the case of West Astoria, we are no longer dealing with a hidden economy existing in the interstices of the regular economy. When an auto repair shop in a residential area services about 100 cars at any moment, as we found in Sunset Park, we are not dealing with a marginal operation that caters to its neighbors. The existing regulatory apparatus in combination with the growing demand for certain goods and services has led to the formation of industrial districts and major servicing areas. But the magnitude of the demand that makes their formation and expansion possible is of another order from that represented by immigrant consumers and entrepreneurs—it is citywide.

In sum, our conceptual elaboration of general economic trends and preliminary findings leads us to posit that the growth in informal activities in New York City is *not* primarily a result of individual survival strategies resulting from unemployment or illegal resident status. Rather, it is primarily an outcome of structural rearrangements in the overall economy to which people are responding. In order to respond, however, individuals must be positioned in certain ways. Immigrant communities represent a "favored" structural location to seize the opportunities for entrepreneurship as well as the most undesirable jobs being engendered by informalization.

CONCLUSION

Over the last two decades, New York City has undergone rather pronounced changes in its economic and social organization. The territorial expression of these changes is most evident in Manhattan, the site of the core economic sector. It is made evident in the expansion of high-rise luxury office buildings and the associated high-income commercial and residential gentrification. These developments have come to incorporate a significantly larger area of Manhattan and adjacent areas of the other boroughs than was the case in the late 1960s. Over this same period, we have seen an expansion in non-European immigrant communities containing a sort of ethnic subeconomy. Along with these two forms of occupation, there also has been an expansion in economically devastated commercial and residential areas.

These spatial arrangements are outcomes of more fundamental processes. But once implanted, they have feedback effects on the spatial organization of production and social reproduction. The purpose of this chapter was to root these spatial arrangements in the sphere of production and in their interaction with the sphere of social reproduction, and, second, to understand to what extent this spatial fragmentation actually incorporates various forms of integration of production and consumption.

One fundamental form of the interaction of space, production, and social reproduction is that the economic base in the city requires a supply of both high-income and low-wage workers. Insofar as it is necessary for a city to contain housing appropriate to both types of workers, conflicts may emerge over the access to city land for housing and to all other components of the sphere of social reproduction—shops, schools, services. At the same time that the economy needs these two types of workers, the bidding power of the high-income workers is so much higher that it can displace spatially the low-income workers. Yet if the trip to work for these low-income workers becomes unacceptably long or costly to them, the economy of the city would suffer, notably those highly dynamic sectors that generate the greatest value and a critical mass of high-income jobs.

The socioeconomic transformation of the resident population has its own "autonomous" effect on space in the city. One effect of expanding high- and low-income populations is the proliferation of small firms; and two very different types of small firms, some catering to the upscale market, others to the low-scale market, and including not only services and retail, but also certain types of manufacturing. Indeed, preliminary data suggest that the existence of high- and low-income residential populations may have the effect of promoting certain forms of manufacturing activity in the city, because, in different ways, each of these population sectors directly or indirectly generates a demand for products and services that are best, or at least suitably, provided through firms that have small scales of production and are local. On the upscale end, these products and services are adequately or conveniently described as customized; on the low scale, as extremely cheap. The proliferation of small firms entails that a city location is feasible and that neighborhood-level firms become more typical at the same time that global market firms dominate the city's economy.

An intermediate variable that is quite significant in facilitating the proliferation of small firms is the existence of immigrant communities because these tend to have the requisite structures to implement such production. These structures include labor supply networks that are both flexible, reliable, and cheap; a supply of skills and entrepreneurial experience—many immigrants having been craft workers in their home countries and holding jobs in immigrant firms here; and mutual help arrangements essential for securing the necessary capital and so on. Spatially, these immigrant communities emerge as key locations for a range of firms that are actually producing for the mainstream market, for select boutiques, for larger firms, and for individual clients.

The overall sense, particularly accentuated in Manhattan, is that residential space has increasingly been occupied by a growing urban gentry and the infrastructure of shops it requires; by a growing number of immigrant communities containing various subeconomies; and by a growing mass of poor, displaced people who occupy devastated areas of the city, some of which may eventually be incorporated into the expanding gentrification process, or, alternatively, the expanding immigrant communities.

NOTES

1. Singelmann and Tienda (1985) found an increase in the extent of intraindustry occupational reorganization during the second half of the 1970s in the United States. They found that the main source of occupational change during this period was not the secular decline in the transformative and extractive industries and the shift to services, but rather changes in job distribution within stable industrial sectors.

2. Among the industries paying under $12,500 in average 1980 wages were: eating and drinking places ($5,330); hotels and lodging places ($7,608); medical services, except hospitals ($8,639); apparel ($9,217); retail trade, except eating and drinking places ($9,969); local transit and intercity buses ($10,140); household furniture ($11,248); real estate ($12,200); business services not elsewhere classified ($12,225); and hospitals ($12,324). Most of these industries are among those generating the largest numbers of new jobs. Also, most of these industries are in the services category. In contrast, most of the industries that paid average wages of $22,000 or more in 1980 are among those losing jobs and they belong mostly to the manufacturing sector. These are mining of nonferrous metal ores ($22,513) and the production of engines, turbines, generators ($22,651); plastic, synthetic rubber ($22,689); metal containers ($22,735); industrial chemicals ($24,121); and blast furnace and basic steel products ($25,496). (See Bluestone et al. 1984: 27.) This classification is based on the Bureau of Labor Statistics codification.

3. For example, in 1984, New York City received 17.2 million tourists according to the Port Authority of New York and New Jersey.

4. These developments lead me to posit that the growth of a middle class and unionization are not only linked through the effects of unionization on jobs (better wages, social benefits, and so on) but also through the effects of a large middle-class on consumption (e.g., demand for standardized, quality-controlled goods, the organization of retail activity into large shopping malls in the suburbs, department stores with many different branches all selling the same kind of product).

5. Here I would disagree with the common notion that the core will specialize in highly technical and professional labor and the periphery in manual labor. The latter may be occurring to a certain extent, but the former is not.

6. Seley (1981) found accessibility to New York City's major markets, suppliers, transport, and labor to be a key factor in making a city location desirable for small businesses, in spite of high costs in a number of inputs, notably energy.

7. Over half of all immigrants admitted from 1950 to 1960 were Europeans. By 1980, almost 80% of all immigrants came from Latin America, the Caribbean, or Asia.

8. See Portes and Walton (1981) for a good discussion of the range of definitions associated with the concept "informal sector." I prefer the notion of "informalization" to describe what is occurring in cities like New York. It is difficult to establish at this point whether we are seeing the development of something akin to a sector.

9. See Portes and Walton (1981) for an examination of the linkages of the informal sector in Third World cities with basic processes of accumulation both at local and international levels.

10. Of interest here is the case of the Netherlands, a country that, according to various socioeconomic indicators of development, ranks among the highest in the developed world. There is growing evidence about various forms of informalization—from industrial homework (Van Luijken, 1983) to underground economic activities (Renooy, Mevissen, and Van Geuns, 1983). Of significance to my hypothesis linking informalization with structural characteristics of advanced industrial economies in the current phase is the fact that much of the economic activities involves native Dutch workers rather than immigrants.

11. From a report issued by Commissioner Fruchter; additional information was obtained from Stuart O'Brian, Inspector General, Buildings Department. According to information from the Union of Construction Contractors, 33% of the $4 billion construction industry is nonunion (1984-1985) with much unregistered work taking place

either directly or indirectly through subcontracting. A recent accident on a mid-Manhattan construction site for a massive new high-rise building led to the discovery that one of the crane operators was a subcontractor and unlicensed. This led to a citywide inspection of major construction sites—that is to say, sites where general inspections are frequent, most labor is unionized and visibility is high. It confirmed the widespread use of subcontractors for jobs that were once done through direct hiring of union labor. The extent to which subcontractors used on such sites were unlicensed took most policymakers and the press by surprise in view of the intense public attention and frequent inspections afforded major sites.

12. From Mr. McGurtle of the General Trade Association for Union Contractors, New York City; confirmed by many other sources (see Sassen-Koob and Grover, 1986).

13. See Balmori (1983).

14. From Paul Grazione, Director of the Enforcement Commission of the Department of Finance and Taxation.

15. Paul O'Brian of the Building Contractors Association of New York City.

16. See Abeles, Schwartz, Hackel & Silverblatt, Inc. (1983); Leichter (1982); New Jersey State Department of Labor (1982); New York State Department of Labor (1982a, 1982b); Sassen-Koob and Benamou (1985); Waldinger (1985).

17. From Mr. Figuaro, Director, Shoe Division of the ACTWU (Amalgamated Clothing and Textile Workers Union).

18. Sassen-Koob and Grover (1986: interview no. 15). Data from the Department of Finance and Taxation show growing violations in footwear manufacturing in the city.

19. Sassen-Koob and Grover (1986: interview no. 3). Conversions of lofts and warehouses into elegant designer residences or shops and galleries has created a large demand for customized woodwork.

20. From Bob Satzger of the Jackson Heights Community Development Corporation.

21. Sassen-Koob and Grover (1986: interview no. 22). Also found was growth of homework in packaging notions, including name brand cosmetics, for high sales periods, e.g., Mother's Day and Christmas.

22. New York State Department of Labor (1982); New Jersey Department of Labor (1982); Sassen-Koob and Benamou (1985).

23. Sassen-Koob and Benamou (1985).

REFERENCES

Abeles, Schwartz, Hackel & Silverblatt, Inc. (1983) The Chinatown Garment Industry Study. New York City: Local 23-25 of the International Ladies Garment Workers Union and the New York Skirt and Sportswear Association.

APPELBAUM, E. (1984) "'High tech' and the structural employment problems of the eighties," in E. Collins (ed.) American Jobs and the Changing Industrial Base. New York: Ballinger.

BALMORI, D. (1983) "Hispanic immigrants in the construction industry: New York City, 1962-1982." Occasional Paper No. 38. New York: New York University, Center for Latin American and Caribbean Studies.

BAUER, D. (1980) "The question of foreign investment." New York Affairs 6: 52-58.

BLUESTONE, B. and B. HARRISON (1982) The Deindustrialization of America. New York: Basic Books.

BLUESTONE, B., B. HARRISON, and L. GORHAM (1984) "Storm clouds on the

horizon: labor market crisis and industrial policy." Boston: Economics Education Project.
City of New York (1986) Report on Economic Conditions in New York City: July-December, 1985. New York: Office of Management and Budget, and Office of Economic Development.
COHEN, R. B. (1981) "The international division of labor, multinational corporations and urban hierarchy," pp. 287-315 in M. Dear and A. J. Scott (eds.) Urbanization and Urban Planning in Capitalist Society. London and New York: Methuen.
Columbia University (1986) Development and Preservation in Manhattan's Chelsea. New York: Columbia University, Program in Urban Planning.
Columbia University (1987) The Informal Economy in Low-Income Communities in New York City. New York: Columbia University, Program in Urban Planning.
Conference Board (1981) Announcements of Foreign Investments in U.S. Manufacturing Industries (First Quarter, Second Quarter and Third Quarter). New York: Author.
Conservation of Human Resources Project (1977) The Corporate Headquarters Complex in New York City. New York: Columbia University, Conservation of Human Resources Project.
DENISON, E. (1979) Accounting for Slower Economic Growth: The U.S. in the 1970's. Washington, DC: Brookings Institution.
DRENNAN, M. (1983) "Local economy and local revenues," pp. 19-44 in M. Hornton and C. Brecher (eds.) Setting Municipal Priorities 1984. New York and London: New York University Press.
FERNANDEZ-KELLY, N. P. with A. M. GARCIA (forthcoming) "Informalization at the core: Hispanic women, home work, and the advanced capitalist state." In A. Portes, M. Castells, and L. Benton (eds.) The Informal Economy: Studies in Advanced and Less Developed Countries. Baltimore, MD: Johns Hopkins Press.
FRIEDMANN, J. and G. WOLFF (1982) "World city formation: an agenda for research and action." International Journal for Urban and Regional Research 6 (3): 309-344.
GINZBERG, E. and G. J. VOJTA (1981) "The service sector of the U.S. economy." Scientific American 244 (March): 48-55.
GUTMANN, P. M. (1979) "Statistical illusions, mistaken policies." Challenge 22 (November-December): 14-17.
HARRIS, R. (1988) "Home and work in New York since 1950." Paper prepared for the Workshop on the Dual City, Committee on New York City, Social Science Research Council, New York, February 26-27.
HOOVER, E. and R. VERNON (1962) Anatomy of a Metropolis. New York: Anchor.
JACOBSON, L. S. (1978) "Earnings losses of workers displaced from manufacturing industries," in W. G. DeWald (ed.) The Impact of International Trade and Investment on Unemployment. Washington, DC: Government Printing Office.
LEICHTER (1982) "Sweatshops to shakedowns: organized crime in New York's garment industry." New York: Office of the 29th District of the State of New York.
MARSHALL, A. (1983) "Immigration in a surplus-worker labor market: the case of New York." Occasional Paper No. 39. New York: New York University, Center for Latin American and Caribbean Studies.
MEYERS, J. (1980) "GNP: perspective on services." New York: Columbia University, Conservation of Human Resources Project.
MORALES, R. and R. MINES (1985) "San Diego's full service restaurants: a view from the back of the house" (report). San Diego: University of California, Center for U.S.-Mexico Studies.

NACLA (North American Congress on Latin America) (1978) Capital's Flight: The Apparel Industry Moves South. Latin America and Empire Report 11 (3, Special Issue).

NASH, J. and N. P. FERNANDEZ-KELLY [eds.] (1983) Women, Men and the International Division of Labor. New York: State University of New York Press.

New Jersey Department of Labor (1982) "Study of industrial homework." Trenton, NJ: Office of Wage and Hour Compliance, Division of Workplace Standards.

New York State Department of Labor (1982a) Report to the Governor and the Legislature on the Garment Manufacturing Industry and Industrial Homework. Albany: Author.

New York State Department of Labor (1982b) Study of State-Federal Employment Standards for Industrial Homeworkers in New York City. Albany: New York State Department of Labor, Division of Labor Standards.

Port Authority of New York and New Jersey (1981) Inventory of Foreign Owned Firms in Manufacturing, by Major Industry in the New York-New Jersey Metropolitan Region. New York: Planning and Development Department, Regional Research Section, Port Authority of New York and New Jersey.

Port Authority of New York and New Jersey (1982) Regional Perspectives: The Regional Economy, 1981 Review, 1982 Outlook. New York: Planning and Development Department, Port Authority of New York and New Jersey.

Port Authority of New York and New Jersey (1985) Economy: 1984. Review: 1985. Outlook for the New York-New Jersey Metropolitan Region. New York: Planning and Development Department, Port Authority of New York and New Jersey.

PORTES, A. (1983) "The informal sector: definition, controversy, and relation to national development." Review, 7 (1): 151-174.

PORTES, A. and J. WALTON (1981) Labor, Class and the International System. New York: Academic Press.

RENOOY, P. H., W. M. MEVISSEN, and R. Van GEUNS (1984) Twilight Economy: A Survey of the Informal Economy in the Netherlands (research report). Amsterdam: University of Amsterdam, Faculty of Economic Sciences.

SAFA, H. I. (1981) "Runaway shops and female employment: the search for cheap labor." Signs 7 (Winter): 418-433.

SASSEN, S. (1988) The Mobility of Labor and Capital. New York and London: Cambridge University Press.

SASSEN-KOOB, S. (1981) "Exporting capital and importing labor: Caribbean migration to New York City." Occasional Paper No. 28. New York: New York University, Center for Latin American and Caribbean Studies.

SASSEN-KOOB, S. (1984a) "The new labor demand in global cities," pp. 139-171 in M. P. Smith (ed.) Cities in Transformation: Capital, Class and Urban Structure. Beverly Hills, CA: Sage.

SASSEN-KOOB, S. (1984b) "Growth and informalization at the core: the case of New York City." Presented at the Symposium on the Informal Sector, Johns Hopkins University, Baltimore, MD, June.

SASSEN-KOOB, S. and C. BENAMOU (1985) "Hispanic women in the garment and electronics industries in the New York metropolitan area." Research Progress Report, presented to the Revson Foundation, New York City.

SASSEN-KOOB, S. and W. GROVER (1986) "Unregistered work in the New York Metropolitan Area" (working paper). New York: Columbia University, Program in Urban Planning.

SELEY, J. E. (1981) "Targeting economic development: an examination of the needs of small businesses." Economic Geography 57 (January): 34-51.

SINGELMANN, J. (1978) From Agriculture to Services: The Transformation of Industrial Employment. Beverly Hills, CA: Sage.

SINGELMANN, J. and M. TIENDA (1985) "The process of occupational change in a service society: the case of the United States, 1960-1980," pp. 48-67 in B. Roberts, R. Finnegan, and D. Gallie (eds.) New Approaches to Economic Life: Economic Restructuring, Unemployment, and the Social Division of Labor. Manchester, England: University of Manchester Press.

SPITZNAS, T. (1981) "Estimating the size of the underground economy in New York City." Regional Economic Digest (a semiannual publication of the New York Regional Economists' Society) 1 (1, Fall).

STANBACK, T. M., Jr., P. J. BEARSE, T. J. NOYELLE, and R. KARASEK (1981) Services: The New Economy. Totowa, NJ: Allanheld, Osmun.

STANBACK, T. M., Jr., and T. J. NOYELLE (1982) Cities in Transition: Changing Job Structures in Atlanta, Denver, Buffalo, Phoenix, Columbus (Ohio), Nashville, Charlotte. Totowa, NJ: Allanheld, Osmun.

STEPICK, A. (1984) "The Haitian informal sector in Miami," in The Urban Informal Sector: Recent Trends in Research and Theory. Conference Proceedings. Baltimore: Johns Hopkins University, Department of Sociology.

TANZI, V. (1982) The Underground Economy in the United States and Abroad, Lexington, MA: D. C. Heath.

U.S. Department of Commerce, Bureau of the Census (selected years) County Business Patterns (New York). Washington, DC: Government Printing Office.

U.S. Department of Commerce, Bureau of the Census (1972) Money Income of Households, Families and Persons in the United States: 1970. Current Population Reports: Series P-60, no. 132. Washington, DC: Government Printing Office.

U.S. Department of Commerce, Bureau of the Census (1982) Money Income of Households, Families and Persons in the United States: 1980. Current Population Reports: Series P-60, no. 132. Washington, DC: Government Printing Office.

U.S. Department of Commerce, Bureau of the Census (1983) 1980 Census of Population: Place of Work, New York. Washington, DC: Government Printing Office.

VAN LUIJKEN, A. (1983) "Thuiswerk." Special issue of Zrij Nederland (October). Amsterdam.

WALDINGER, R. (1985) "Immigration and industrial change in the New York City apparel industry," pp. 323-349 in G. Borjas and M. Tienda (eds.) Hispanics in the U.S. Economy. New York: Academic Press.

WANG, J. (1979) "Behind the boom: power and economics in Chinatown." New York Affairs 5: 3.

WOLFF, G. (1984) "The decline of unionization and the growth of employment in the Los Angeles region." Los Angeles: University of California, Graduate School of Architecture and Urban Planning.

4

Industrial Restructuring and Regional Politics

ANN ROELL MARKUSEN

REGIONAL ECONOMIES in the United States remain highly differentiated, despite decades of convergence in per capita incomes and heightened capital mobility. The map of the United States is densely dotted with "enclave economies"—industry-specific production units that dominate the economic lives of their surrounding communities. At the regional level, economies are marked by quite disparate industrial structures, growth dynamics, and inherited management-labor relations systems.

In this chapter, I show how these different industrial situations produce distinctive brands of regional politics. Using the profit cycle model of industrial evolution, I distinguish between three general types of enclave economies: those booming from indigenous expansion of new, innovative superprofit sectors; those expanding with an influx of older, mature profit-squeezed sectors; and those contracting from the exit of those same profit-squeezed sectors. I then show how localities and regions enjoying a superprofit boom will develop a regional politics pivoting on internal conflicts over land use, infrastructure, and environmental issues. Similarly, regional economies on the receiving end of migrating profit-squeezed sectors will evolve internally conflictual politics, but the substantive issues will revolve around preserving the business climate—low wages, absence of unions, a low social wage—which is the basis for their newfound prosperity. In contrast, regional economies besieged with the loss of traditionally strong sectors will tend to develop an externally oriented politics that levies claims upon the federal state for redistribution and redress.

In each case, a number of examples are offered at both the local and the regional levels. I contrast three emergent regional coalitions of the 1970s—the Southern Growth Policies Board, the Western Governors' Policy Office, and the Northeast-Midwest Congressional Coalition—and show how their politics were shaped by the underlying industrial restructuring process. In that decade, these groups assumed for a time quite antagonistic postures vis-à-vis federal government policy. Thus differentials in regional industrial growth patterns can produce regional conflict. I show that the potential for such conflict depends upon a number of preexisting conditions within and among the regions involved.[1]

INDUSTRIAL RESTRUCTURING AND REGIONAL ECONOMIES

What distinguishes regions and subregions from each other are the relatively unique sectoral configurations of their local economies. Regions and cities manifesting strong regional consciousness and a distinctive regional politics often owe their predicament to one or more sectors that dominate the local economy.

We begin, then, with a characterization of industrial dynamics. Individual sectors or industries (e.g., aluminum, autos, brewing, chemicals) display quite distinct forms of competition and corporate strategy at sequential stages in their development.[2] Most sectors pass through a life cycle across which the conditions governing profitability vary dramatically. Production and marketing decisions, including where to locate plants and jobs, evolve in a relatively predictable manner as the source and size of profits change. In a formulation I call the profit cycle, several stages can be identified, three of which are most apt to ignite new forms of regional politics: an innovative stage of superprofits, a competitive stage of "normal profits," and a hypercompetitive stage of profit squeeze.

INNOVATION, SUPERPROFITS, AND AGGLOMERATION

When a new industry is "born," as, for instance, the steel industry in the 1870s or the electronics industry in the 1960s, it generally enjoys an initial era of extraordinary growth and high returns. These *superprofits*, defined as the difference of revenues over cost per unit, arise from the special use value that the new commodity has in contrast to existing methods or products.[3] For a time, potential competitors will have a difficult time entering the new market, because of lack of expertise, delays in organizational formation, and the discouraging presence of patents. Superprofits will not necessarily make original entrepreneurs rich, but are generally plowed back into new capital expenditures and experimentation to expand the business and to maintain market leadership.

In such an innovative era, the firms forming a new sector will tend to cluster around one or a few nodes. These sites may be close to existing pools of professional/technical labor and corporate headquarters, or they may be in quite accidental locations (an inventor's or founder's residence).[4] Wherever this initial siting, the rapid growth of the industry in its youth will tend to reinforce the importance of these sites as agglomerative centers.[5] New firms are frequently formed from employees spinning off from older firms.

In these new industrial enclaves, competitive efforts will focus on product design, on prototype construction and testing, and on organization building—all activities that rely heavily upon a skilled labor pool, skewed toward the professional technical categories. The young firms will tend to recruit labor from outside the region and to attract subcontracting firms and suppliers around them. Any inclination to disperse production to cheaper peripheral sites will be tempered by the need to be near the center of ongoing innovative activity and to have ready access to new information about the evolving market. Shoemaking in the Boston area is an early example of this type of innovative agglomeration; autos in Detroit, rubber in Akron, steel in Pittsburgh, farm machinery in Chicago, brewing in Milwaukee, flour milling in Minneapolis, oil in Houston, cereals in Battle Creek, and electronics in Santa Clara Valley are others.

STANDARDIZATION, NORMAL PROFITS, AND DISPERSION

As firms successfully perfect a product line and set in place a standard process for large-scale production, much of the strategic need to be near the centers of origin disappears. Organizational attention turns toward market penetration and away from the innovative process itself. The superprofits of the previous period disappear under heightened competition from new entrants, and the major source of expanded profits thus becomes an increase in the volume of sales to capture economies of scale and spread organizational overhead.[6] At the same time, firms in this stage require fewer experimental and innovative personnel and have less interest in remaining close to competitors.

In this stage, corporations will find it more attractive to site new plants at least-cost locations and near developing markets. Textile plants have migrated to the American South and offshore in the past 50 years because labor costs at these sites are so much lower; Coca-Cola bottling plants have been located in every major metropolitan area. In some cases, higher costs at the initial location—through labor organizations, and through pressures on locally scarce resources like land and air quality—will help push production out. In general, then, industries in this stage will tend to disperse from original sites under the discipline of normal competition.[7]

MARKET DECLINE, PROFIT SQUEEZE, AND PLANT CLOSINGS

A final era is characterized by shrinkage of output and the decline of profit rates to a less-than-average level.[8] Most commodities in capitalist society face ultimate extinction as new substitutes emerge to replace them—a function of the relentless search for new profit arenas that is endemic to capitalist systems. Take, for example, copper. Most major U.S. mines have shut down in recent years, in part due to depletion and to import penetration, but in large part due to the substitution of fiber optics in communications uses, of polyvinylchloride in sewer and water pipes, and of aluminum in radiators. If such new substitutes are not pioneered internally, the industry producing the older commodity will find its market demand dropping and its profit rate falling. It will respond by trying to rationalize—closing down older capacity, and if entrepreneurial, modernizing existing operations and diversifying into new product lines.

Generally, the net effect of this behavior is the systematic elimination of capacity in the older regions, with maintenance or at least fewer closings in regions of more recent entry. As a result, the industry will appear to be shifting spatially, even reconcentrating its capacity at the more youthful sites. Take the textile and shoe industries, for example. Once almost entirely a New England phenomenon, newer plants were dispersed to lower-cost locations in the southeastern Piedmont and the Kentucky/Tennessee interior throughout the first half of the twentieth century. Once heightened competition from imports pressed domestic producers, the older New England plants were eliminated, so that both of these sectors now seem to have reconcentrated in a small number of southeastern states.[9]

ASSOCIATED REGIONAL DEVELOPMENT PATTERNS

These three types of profitability experience, generally found sequentially, and their behavioral counterparts, are quite helpful for distinguishing among regional growth dilemmas. They help explain why, even in an era of rapid market interpenetration and huge wage gaps across countries, some regions of the United States are still hosting job growth. Indeed, this profit cycle model enables us to distinguish between two very different forms of regional growth. A large increase in employment registered in a region may mean either the birth of a new sector or the dispersion or relocation of a more established sector from some other region.

The local consequences of these two types of sectoral growth are dramatically different. In the former, we might expect the generation of a highly skilled labor force and associated agglomerative activity, which can set off a major long growth spurt. In the latter case, long-term growth prospects may be an illusion, as the relocation is due solely to the response of corporations to the relative underdevelopment of the regional economy. As this new presence helps create a local labor market, pushes up wages, and encourages unionization, these very benefits may choke off any further growth. The same cost-cutting insistence that encouraged the original relocation may lead to a further round of relocations to offshore sites, rendering the promise of sustained regional development even more ephemeral.

These distinctions between stage status of incoming sectors have dramatic consequences for regional development potential. If the new activities are in the superprofit stage, they introduce the in-migration of technical and professional labor, dramatic expansion in local business services, and pressures on housing, transportation infrastructure, and air quality—in other words, they bring robust job growth rates with disruptive land-use and public sector consequences. If the incoming sector is a mature, profit-squeezed activity fleeing high tax and wage levels and lured by financial and other incentives, it will bring to its new host problems similar to those of the region it left—heightened capital-labor conflict and the threat of closure or overseas relocation—as well as pressures on local carrying capacity. Thus the Deep South's efforts to attract post-normal-profit-stage corporations will result in a very different developmental path for the region than will the Southwest's defense and electronics-induced growth.

In addition to aggregate employment levels, the spatial behavior of sectors at different points of the product cycle affects both the occupational composition and the structures of the business community within regions. Firms in the initial, superprofit stage will tend to have relatively high proportions of professional/technical workers. As the production work force grows subsequently, a dual work force emerges in the region.[10] Corporations in the last stage will tend also to have a bifurcated class structure—managerial and production—but it is more likely to be spatially segregated inter- or intraregionally. The growth of employment, therefore, will result in very different occupational structures locally, depending on which stage—agglomeration or relocation—of the profit cycle is involved.

Entrepreneurial and unionization structures at the local or regional levels will also be affected by sectoral type and dynamic. Regions with dominant sectors in the early stages of development will tend to have a large and robust entrepreneurial class composed of managers and proprietors of small firms. Regions hosting new dispersals or relocations from existing corporations in other regions will experience an absentee capitalist class, quite distinct from local small business persons in the tertiary sector. Unionization will not typically be extensive in these locations, although efforts to organize workers will sooner or later develop.

In summary, then, the determinants of profitability for sectors at different developmental stages will create differential locational impulses. Superprofit sectors will agglomerate, creating pressures for new infrastructure and amenities required for their professional-stratum employees. Profit-squeezed sectors will disperse in search of lower production costs and more efficient market penetration. Three distinct regional experiences are linked to the profit cycle's rhythm: regions losing profit-squeezed sectors, regions gaining the same, and regions hosting youthful, superprofitable sectors. Each, as we shall see, is associated with specific regional political tendencies.

TRANSLATION INTO REGIONAL POLITICS

Regional politics often emerge from these dynamics of individual economic sectors and their associated enclaves. Sectors in specific stages of the profit cycle manifest tendencies toward certain forms of regional politics. These vary by the nature of the issues raised, the protagonist members of a coalition, and the level of the state to which the demands are addressed. If two regions with a predominance of sectors in dissimilar stages of the profit cycle coexist in one nation, then regional antagonisms over federal government policy may emerge. In particular,

TABLE 4.1
Regional Coalition Membership, Issues, and Unity Levels Associated with Profit Cycle Stages

Stages	*Coalition Members and Opponents*	*Major Issues*	*Level of Government Most Involved*	*Degree of Internal Unity Within Region*
Superprofit boom region	Incoming corporations, progrowth coalition versus residentialists, traditional sectors, business interests linked to traditional sectors	Land use, infrastructure; environment	1. Local 2. State 3. National	Low
Region of entry	Regional business, business	Maintenance of low wages, low taxes; growth management; unionization	1. State 2. Local 3. National	High
Region of exit	Workers, local business versus transregional corporations, other regions	Reinvestment, capital controls, compensation, adjustment aid	1. National 2. State 3. Local	High

sectors in two distinct phases of the profit cycle may fuel such regional conflicts. The two correspond to the rapid growth and rapid decline stages of an industry's evolution. At either of these points in time, extraordinary changes in employment (up or down), heavy demands upon resources, and severe multiplier effects on local businesses and residents may culminate in a new form of regional politics, either internally or supraregionally. The major elements of this model of regional politics are tabulated in Table 4.1. Throughout the following discussion, examples from recent U.S. history will be drawn upon to illustrate the argument.

The relationships between economic structure and regional politics hypothesized here are not meant to assert that the former is the sole, nor even the primary, determinant of regional politics. Regional and local cultures are quite significant ingredients in the formation of regional politics. The presence or absence of a unifying culture, a tradition of militance, or a prevailing ideology for or against private enterprise can be a determining factor in the generation or outcome of a particular conflict.[11] Furthermore, not all cases of economic stress will provoke regional conflict. Some conflicts will be fought out solely within sectors, between labor and capital, or among factions of both, and some will never produce any form of conflict at all. But economic differentiation and change are important, often dominant, elements in most regional struggles, and they produce generalizable tendencies toward specific forms of regional politics.[12] The framework presented here is designed primarily to trace the specifically economic impulses to regionalism.

Second, this analysis of regionalism is specifically tailored to the contemporary United States. Modern American political structure is a particularly important contextual factor. Because regionalism consists of a set of territorial claims levied on the state, its possible forms and content depend upon the structure of that state and what kinds of territorial politics it currently permits. In the United States, the three-tiered nature of the state, with its federal, state, and local levels, is a primary element assumed in the treatment. Crucial aspects for postwar regional politics include the following: intergovernmental boundaries; intergovernmental allocation of responsibilities and resources; functional rules such as representative democracy; the division of powers among the executive, legislature, and courts; the dual commitment to private property rights and civil liberties; and the two-party system. In what follows, these various elements will be scrutinized when they are central to the interpretation.[13]

THE POLITICS OF BOOM ECONOMIES

Sectors in innovative, superprofit stages will tend to produce predominantly intraregional and local planning conflicts. The primary struggles will surround the transfer and use of local resources. First of all, firms in the new sectors are apt to bid up the prices for land, labor, and water as they put pressure on an existing stock of these resources. Second, a derived pressure on local resources will operate through the expanded work force, if in-migration accompanies the new sectors' growth, exacerbating the direct growth effects of the sector itself. The housing stock may become inadequate and, therefore, prohibitively expensive. The transportation system may become overloaded, causing significant delays, wasting commuting time, and increasing air pollution. In this atmosphere, in an unrestricted capitalist system, the pricing mechanism will tend to shift resources from prior users to new sectors. For those who own the resources, this may enhance their wealth, so they may be expected to favor the entry of the new sector. For those sectors and subsistence users whose viability requires access to these resources at their previously lower price, and for those who worked in such sectors, the growth or entry of new sectors may be viewed with hostility.

The state becomes the forum for confrontation and arbitration of these issues for two reasons. First of all, it makes decisions regarding the use of land, through regulation of property transfers, permission to change land uses, and permission to build. Second, it makes decisions regarding the public investment that is required for an expanding sector to be viable.[14] Expansion generally requires significant new investment in roads, water, and sewerage, expanded protection services, and new commercial and residential infrastructure. Because both these functions are local responsibilities in the United States, struggle around the transfer and use of resources tends to organize itself around the local level. Thus, in initial profit cycle stages, regional politics tends to be intraregional, focused on land-use and public investment issues, and composed of progrowth and slow- or antigrowth coalitions with a distinct sectoral character.

Two exceptions to this pattern may be mentioned at the outset. First, to the extent that the incoming sector is so large that its impact overreaches local jurisdictional boundaries, it will tend to push the conflict up to the next-highest level, in this case, the state. If the supralocal impact is differentiated across units, then conflict between the local governments involved may emerge as well, because each may have different dominant coalitions. An example is the energy boomtown phenomenon in which one town hosts the tax-base-enhancing plant or mine while another is burdened with the housing, school, and public service demands of a growing population of the plant's workers. In cases where jurisdictions suffer the same impact, they may join together in a statewide coalition, or they may still differ about the course of action, depending upon the importance of the particular sector within the region or on the class composition of the community.

The second exception is the case in which the federal government owns a significant share of the land whose future is being debated. This is a critical factor in conflict over western U.S. energy development. Because the federal government is a major, sometimes the sole, landowner, it governs the land-use issue—who gets it, the terms of using the land (lease stipulations), the price of that use (royalties and rent), and the share that the local jurisdiction will recoup. When federal land is involved, regional conflict around changes in the agricultural, energy, and tourism sectors necessarily involves conflict at the federal, state, and local levels.

THREE LOCAL EXAMPLES FROM ELECTRONICS, ENERGY, AND DEFENSE

Three quite disparate examples of this type of sectorally based regional politics can be offered as examples. First, the combination of fast-growing aerospace, electronics, computer, and software sectors in the Santa Clara Valley south of San Francisco has set off a process of dramatic competition for land and human resources. The price of land has been bid up so high that firms are increasing work-space densities by building upward and putting two workers in a space previously occupied by one. The lure of the area's concentration of professional/technical jobs has drawn many well-educated in-migrants who have bid up the price of housing, causing particularly severe problems for lower-income groups. The area's transportation system has become clogged as workers take on longer and longer commutes. The quality of the air suffers as a consequence. The net result has been a strong residentialist movement to limit the addition of new industrial space—the first example of industrial growth controls in the country. It has also spawned a countermovement by real estate and industrial interests to eliminate environmental and community controls that hamper the construction of new housing (see Saxenian, 1980).

A second example is the western energy boomtown case, perhaps best exemplified by the remarkable events in Southeastern Montana coalfields in the mid-1970s. Here, extensive strip mining and accompanying coal-fired electricity plants raised significant opposition from ranchers, Native Americans, and old-time members of the community. Issues included land use (especially federal leasing provisions), air quality (a landmark court case stopped construction on two major power plant units), financing of local infrastructure (solved with a 30% severance tax on coal mined), and control of development (met with statewide siting legislation). Both interjurisdiction disputes and federal land ownership helped carry the issues from the local level to the state and federal levels.

A third set of examples, less well documented but clearly emerging, are the conflicts in areas where military-related production brings a boom to local economies. One of the factors underlying the city of Santa

Monica's strong rent control movement is the upward pressure on housing prices in west Los Angeles Basin communities caused by the burgeoning defense and space-related complexes in El Segundo and other nearby communities. Similarly, opposition to developers' carte blanche is growing in cities like Colorado Springs, as a similar boom propels the rapid conversion of the high plains into sprawling bedroom communities for defense and space-related branch plants. The most powerful recent instance was the Utah and Nevada coalition of ranchers, environmentalists, and the Mormon church that opposed the MX missile in the early 1980s.[15]

A REGIONWIDE COUNTERPART: THE WESTERN GOVERNORS' POLICY OFFICE

One such sectorally preoccupied case of boom regionalism appeared in the intermountain West in the 1970s. The states rimmed by the Rockies on the east and the Sierras on the west have always been an energy, precious metals, and agricultural hinterland to the rest of the nation. A region with a powerful ideology of rugged independence, resource exploitation has paradoxically rendered it more heavily dependent upon outside demand and capital than most. The western economy has several enduring features that distinguish it from the rest of the nation: poor internal linkages, severe intersectoral competition for scarce natural resources, substantial outside ownership of land and productive capacity, and an extraordinary degree of federal government land management.

For decades, most of the region was stagnant; out-migration rates reached as high as 30% in many counties, especially in the 1950s. Into this sparsely settled economy came the energy boom of the 1970s. Altogether, the region was slated to host 250 major new coal mines, 150 new coal-fired power plants, and 20 major new synthetic fuels plants, plus pipelines, transmission lines, railroad spurs, and new roads to transport the energy out (Baldwin, 1973). While not entirely new industries, the activities initiated—high-tech strip mining of coal, experiments to crush oil out of shale, and expensive new synthetic fuels plants—were quite innovative, even speculative, and generated substantial superprofits for their backers, at least for a time. Many of the features of a youthful agglomerating industry accompanied the energy boom: smaller communities swamped with construction activity and newcomers; heightened conflict over land use and infrastructure; and booming core cities like Denver, where the energy companies' offices and phalanx of geologists were concentrated.

A number of adversely affected diverse groups mounted a challenge to the incoming energy companies and what they saw as their federal government handmaidens. In two early victories, the environmentalists stopped the 1976 Winter Olympics in Colorado, which threatened a pristine recreational environment, and the Kaiparowitz power plant, which threatened air quality in southeastern Utah. Ranching and farming communities mobilized to oppose strip mining, coal-fired plants, and high-voltage transmission lines, which intruded upon their land, water supply, air quality, and settled community way of life. Explosive boomtown growth led many mountain towns to fight associated fiscal pressures and community disruption. Native Americans contested the leasing terms on their energy-rich land, the inadequate royalties, and the ineffective employment policies.[16]

These disparate disgruntled citizens elected new-style politicians in the early 1970s, particularly Governor Thomas Judge in Montana and Governor Richard Lamm of Colorado. Both were unabashed environmentalists and anticorporatists, as this press statement by Judge in October 1974 shows:

> Decisions should be made by the people of this region, not by the federal government or the energy companies. If we sit passively by, we will be a colony for the rest of the United States. . . . We've got to organize or the Eastern banks and corporations, helped by the Ford administration, will pick us off one by one [Hayes, 1980: 3].

By 1976, these two had organized the rest of their intermountain counterparts into the Western Governors' Regional Energy Policy Office, with goals on two fronts. The first was an internal effort to unify states around a strong growth control stance and to strengthen the ability of each to deal with large energy companies. Among the principal demands were stiff severance taxes both to deter energy exploitation and to pay for the associated costs of community disruption, and tough siting regulations that would give the governors control over where, when, and how new facilities and mines were installed. The second was external—the effort to present a united front to the federal government, whose policies determined mineral leasing, land-use, water, and infrastructure allocations (Hayes, 1980: 79, 96-97, 107; Peirce and Hagstrom, 1977: 208-219).

The governors' coalition, which became the Western Governors' Policy Office (WESTPO) in 1977, accomplished much in its initial years. It successfully protected severance taxes and siting legislation from being dismantled. It rebuffed assaults by the Northeast-Midwest Institute on its energy taxation and regulation innovations. It eventually succumbed to co-optation by the region's utilities and business elites, however, especially as money from the energy boom began to circulate through the economy. Gradually, it shifted its focus away from environmentalism and land-use preservation toward outright enthusiasm for energy development. By 1981, WESTPO was promoting coal exports, with a task force headed up by Tom Judge. Nevertheless, its evolution had provided an arena in which ongoing alternative organizations could argue their causes and gain political access and skills.[17]

This dramatic movement is a good example of emergent politics in a region booming with incoming superprofit sectors. Arrayed against the incoming coal and oil companies were boomtown residents, environmentalists, ranchers, farmers, and indigenous businesses linked to these traditional sectors. The major issues throughout concerned land use, infrastructure, environment, and housing. Local governments were often thrown into political chaos by residents' demands on one side and the enormous revenues proffered by the energy companies on the other. State governments found themselves heavily petitioned by these constituents to employ their considerable power in defense of existing communities. Claims were also levied against Washington, but principally because it was a major landowner and stakeholder in the region. The degree of internal unity was low, which became clear in the ease with which energy companies found indigenous collaborators and redirected the coalition almost 180 degrees.

THE POLITICS OF BUST ECONOMIES

Sectors in mature and declining stages are apt to produce intraregional unity and interregional conflict in the region of origin. The tendency toward internal unity arises from the damaging effect that one or more sectors' decline has on an entire local or regional economy, through the depressing effect of its payroll withdrawal on other regional businesses and on the budgets of local and state governments. In this case, it is possible to win the adherence of local businesses, especially those whose assets and markets are unalterably fixed locally. Churches, local banks with lots of mortgage money tied up in local residential and commercial real estate, real estate firms, local insurance salesmen, newspapers whose circulation is purely local—all have an incentive to join ranks with workers displaced from that sector, as well as other local suppliers who sold only to the shutdown sector, such as janitorial services, parts makers, guard services, accountants, and lawyers. Local unity is further abetted by the fact that the immediately offending party is often an outside corporation or conglomerate. This fact tends to generate a form of nativism, where the exiting corporation, rather than the capitalist development process, becomes the villain.[18]

If sectoral decline or sectorally related issues are so severe that the entire economic base of a region is threatened, a cross-ethnic working-class coalition may gain control of a government previously in the hands of the region's dominant but exiting or absent capitalists. Examples are successful political challenges at the regional level by iron ore miners on the Mesabi Range in the early twentieth century, by gold miners in the Cripple Creek district of Colorado in the same era, and at the state level by copper workers and farmers in Montana in the 1970s. Generally, however, pure cases of working-class control are hard to come by, because some ensemble of business interests almost always complicates regional politics.

Sectors in profit-squeezed stages generate problems for their home regions that readily prompt demands for a policy or planning response at the national level. In contrast, newer, superprofit sectors create pressures that are more often fought out within the region. There are several reasons for this. In the case of out-migration of capital, or sudden plant closings, money capital is not operating within the region, but outside of it. Thus the only resort for a region with a coalition wishing to oppose the out-migration of a sector is to raise the issue in the state forum that governs the larger territory across which sectoral change is occurring.[19]

In the case of newer sectors, the money capital bidding for resources is entering, rather than leaving, the region. Thus the terms of its entry are more apt to be written within the region. If an oppositional group succeeds in deploying some mechanism of the local state, such as zoning, to slow its entry or expansion, the growing sector is more apt to possess the resources and motivation to compromise, compensate, buy off, and restructure its plans. This was particularly true in the case of the energy-rich West, discussed above.

Second, the conventional mechanisms for intervening in or softening the pace and intensity of capital migration are mostly not tools of local government. Local tax concessions have been employed to try to convince corporations not to move or to rebuild locally, as in the recent case of GM and Detroit. But most routes require powers and resources not available to the local government. Most cost-reducing measures, for instance, such as better transportation infrastructure or a better-trained labor force, are the domain of the bureaucracy of state governments. Measures aimed at stimulating demand, such as import protection, are the province of the national government. Even the power to regulate bargaining between labor and management, another route to dealing with decline, resides at the federal level. Thus the traditional division of powers renders state mechanisms for response to the case of decline more inaccessible than in the growth case.

Third, the complexion of local politics differs sharply in the two cases. In the case of a newly emergent sector, we can expect that those with a base in the older sectors may have a strong hold on the machinery of local government, so that their chances of successful protection of their interests are better. Ranchers in western boomtowns are a good example. In the case of an out-migrating sector, the local government is more likely to have been a collaborator with that sector, through campaign financing, local business ties, old family connections; thus a priori it is more difficult for a coalition of the displaced to win a vigorous partisanship from local politicians and their bureaucracies. Unless an unusual turnover has liberalized the prevailing administration and changed the course of its politics (which was true at the state level in Montana), a coalition around plant closings may seek to pursue its cause at higher levels of government, where its success, of course, depends on its antagonists' power at that level as well.

Finally, more radical responses to sectoral decline (such as plant takeovers, or economic revitalization) require massive amounts of capital, not only initially to purchase the physical plant vacated by the outgoing sectors but also to refurbish or modernize it. Such funds are not generally available at the local level, particularly when the region is suffering the rounds of economic base shrinkage that follow a sector's wane. In Detroit, an early 1980s proposal to reshape the area's economy on new metal-bending sectors with greater expansionary promise would have committed a large chunk of federal resources to the transition (see Luria and Russell, 1981).

YOUNGSTOWN: A PROTOTYPE

The Youngstown case in the late 1970s is an outstanding local illustration. When two major steel companies announced the shutdown of multiple facilities within the short span of two years, a broad-based worker/religious/small-business coalition was formed. Its object was to reopen the mills, if possible. It forged substantial internal unity, in large part by critiquing the behavior of outside corporations. They correctly understood that, to achieve the levels of reinvestment required, they would have to carry their case to the federal government. In the end, the effort failed, both because the steel industry had close ties to the Carter administration and because federal grants to alternative economic development enterprises undercut the business communities' support for the coalition. A more sophisticated follow on, however, has emerged in Pittsburgh's Steel Valley Authority initiative (see Lynd, 1982; Tri-State Conference on Steel, 1984).

Once escalated to the national level, the politics of regionalism around declining sectors tends to generate two kinds of conflict over the use of the national government to bolster that region's economy. First of all, the sector whose exit is being protested will generally argue its case through existing bureaucratic and political links. Thus a conflict between workers and the community on the one hand, and the corporations within that sector on the other, emerges at the national level within forums of the state. The balance of this struggle depends upon the support that each contestant gets from other groups in the society. While other corporations will generally support the right of corporations in any particular sector to move, occasional cases may emerge where other sectors support the challenge because of potential adverse effects for them as suppliers or customers. Regional railroads, for instance, may favor the retention of steel-making capacity or textile firms in the Northeast; northeastern regional utilities would prefer that coal mining continue in Appalachia versus the West. A well-developed interregional intersectoral coalition around plant closings has not yet developed, but might in the future.

A REGIONWIDE COUNTERPART: THE NORTHEAST IN DISTRESS IN THE 1970s

Throughout the postwar period, the traditional heartland of the United States, stretching from New England through Baltimore, St. Louis, and Milwaukee, experienced rates of growth considerably slower than the rest of the nation. In the postwar era of accelerating integration of the world economy, this region found itself increasingly unable to maintain its industrial base. Lower transoceanic transportation costs and rapid but low-wage westernization in Third World countries enticed profit-seeking corporations to set up plants abroad. The conglomerate reform of the corporation and the increasing internationalization of world capital markets permitted both capital and commodities greater ease in traversing national boundaries. The Northeast, with its sophisticated, unionized work force and relatively high social wage, was at a growing disadvantage in this more integrated world economy.

In tandem with these general forces, a significant number of sectors in which this region had specialized for a century reached profit-squeeze and abandonment stages of their profit cycles. The industrial blight moved from East to West over three decades. First New England, in the 1950s and 1960s, suffered permanent losses in textiles, shoes, fishing, metals, apparel, and food products. In the 1970s, the Middle Atlantic states were worst hit, with severe shrinkage in apparel, machinery, and metals. These states lost almost 5% of their population to out-migration in the 1970s. By the 1980s, the worst losses were concentrated in the midwestern states of Ohio, Michigan, Illinois, Indiana, Wisconsin, and Iowa, where the machinery and metals sectors lost in excess of 25% of their jobs. In the first four years of the 1980s, out-migration from these states was 3%.[20]

Few new manufacturing sectors moved in to compensate for these losses, except for the heavily military-dependent high-tech industries like jet engines, nuclear submarines, communications equipment, and electronics in lower New England and Long Island. But as the traditional center of American business activity, with the overwhelming number of corporate headquarters and financial markets, the region found itself increasingly bifurcated between financial, military, and business service-related growth in a few centers like Boston, New York, Pittsburgh, and Chicago, while people in the countryside and the industrial cities like Buffalo, Detroit, Akron, and Cleveland found it difficult to sustain their traditional ways of life.

Because the region continues to be home to the majority of U.S.-based multinational corporations, an indigenous capitalist class remains very active in its politics. On the one hand, financial and industrial interests have a tremendous stake in the fixed capital assets of the region. On the other hand, they are simultaneously making decisions to invest money and physical capital outside the region, contributing to its fiscal problems and structural decline. This contradiction explains in part why the leadership of the regional coalition that emerged in the 1970s found it so opportune to point to villains outside the region—the Sunbelt, which was "robbing its jobs," and the federal government, which was taxing the Northeast to pay for growth in the Sunbelt (Havemann et al., 1976; "Jobs and Revenues Shift," 1976; and "A Second War Between the States," 1976).

To tackle the new regional problems of urban fiscal crisis and growing economic distress, a number of governors attempted to form a regionwide coalition, particularly Governor Carey of New York. These efforts were thwarted, however, by the strong partisan politics dominating the region. Republican governors, particularly in the Midwest, refused to follow the lead of their Democratic counterparts farther east. Instead, the Northeast-Midwest Congressional Coalition initiated by Congressman Harrington of Massachusetts became the chief organizing vehicle for the region's assault on the federal government and its Sunbelt and western counterparts.[21] Resources within the region were already hard-pressed, and thus Congress appeared to be the logical arena in which to pursue redress. It is symptomatic of the region's external orientation in this time of trouble that the seat of its organization was the nation's capital, unlike their counterpart regional organizations in the same period.

The coalition and its think-tank arm, the Northeast-Midwest Institute, aggressively produced studies and legislation designed to redirect federal resources into the region. It achieved several formula changes that modestly boosted the level of intergovernmental aid flowing to states and cities in the Northeast. It attacked the role of defense spending in favoring the Sunbelt. It accused the western governors of forming a second OPEC with their severance taxes on energy (Northeast-Midwest Institute, 1978, 1980). Unfortunately, shifting geopolitics undermined the voting clout of the coalition in Congress, so that, by 1980, it was unable to achieve easily its regionalist demands.

The external orientation of the northeast coalition in the 1970s can be seen not only as a response to sectoral deterioration, but to the maturity of class conflict in the region as well. The politicians, chiefly Democrats, that built the coalition owed their success to a delicately engineered base that included both business and working-class constituents. On the one hand, they could not offend their business patrons by talking about effective controls on the movement of capital or novel forms of public investment. On the other, they could not alienate their working-class base by espousing an erosion of the social wage and workers' rights in order to improve the region's business climate. The coalition thus found it attractive to blame outsiders and focus its energies away from the internally conflictual issues of business cost and social control of investment toward the common demand for larger cash infusions, fairer treatment, and additional social capital from the federal level. It stressed the abused position of the region within the nation rather than addressing the opposing interests of capital and labor at home. And because it actually achieved little, concretely, in its forays against other regions and the nation, perhaps its greatest accomplishment was precisely in diffusing the potential internal conflict and achieving, for a time, cross-class unity.

THE POLITICS OF HYBRID ECONOMIES

Regions into which mature sectors are relocating will tend to display a hybrid politics that incorporates elements of each of the two major types just presented. As growth areas, problems of land and resource use will tend to preoccupy them internally. When growth management issues spill over local or state boundaries, they will create pressure for regionwide responses. On these questions, internal regional unity will be strong, led by the indigenous business community. On the other hand, growth management generates significant public sector expenditure demands and erodes the low tax rate that drew relocating firms in the first place. Internal conflict will be greatest around economic features that make the region attractive to profit-squeezed sectors—the low cost of labor (and the absence of unionization), the low cost of land, and the low level of public sector liability. A business-dominated coalition can be expected to make maintenance of its good "business climate" a central regional planning goal.

Conflict between two regions who are competing over the same mature sectors—one as original host and the other as recent host—is apt to emerge over questions of interregional compensation and aid. Opposition by the latter to national petitions by the former for reinvestment funds, indemnities, controls on capital movement, and amplified aid may be joined by regions enjoying expansion in innovative sectors. Both types of growth regions will be unsympathetic to demands for channeling public resources to declining regions, on the grounds that the same national resources could better be used to treat their growth problems. This interregional conflict will be exacerbated if the newer, expanding regions are generally relatively underdeveloped; that is, if they have lower standards of living, less-developed public sectors, and a higher incidence of poverty.

Such was the case in the 1970s with the formation of the Southern Growth Policies Board and its subsequent squabbles with the Northeast-Midwest Institute. In the postwar period, the Old South turned New

South. It is a premier example of a hybrid region. In both the 1950s and 1960s, the region's employment grew one and a half times the national rate. Even in the 1970s, the region posted four new jobs for every three nationally, and in the 1980s, it was relatively recession-proof. Much of this growth, however, was composed of older manufacturing sectors shifting away from locations in other regions. In a world of increasingly competitive Third World economies, where the South is flanked by even lower-cost regions, the development promise of these new industries was ambiguous.

The rapid industrial growth of the South was in large part a catching up of this region to the rest of the country, as the economic legacies of slavery and cash crop agriculture finally loosened their hold on southern institutions and labor markets.[22] For one thing, the mechanization of cotton picking in the late 1940s eliminated much of the demand for black agricultural labor that had kept southern elites committed to a system of discrimination against blacks in industrial employment. For another, the civil rights movement successfully challenged the restrictions on blacks' access to jobs and education. As the labor force swelled, employers in search of a cheap and moderately skilled labor force found the South an amenable location for relocated plants or expansions.

Southern growth was aided by federal largess, as well. The TVA project, begun in the 1930s, improved both commercial agriculture and provided vast new accretions of cheap electric power, engendering manufacturing growth. In addition, large shares of federal military expenditures on military bases and personnel went to southern jurisdictions, prompting economic development in their environs.

Rapid growth in many locations brought to the South the problems of the industrialized North and the energy-booming West—conflicts over land use and resources, as urban sprawl and central-city decline, water shortages, and environmental degradation began to surface as serious problems. But these growth dilemmas were unlike those experienced in Silicon Valley or northwestern Colorado, because they accompanied the entry and growth of sectors in the latter stages of their profit cycles, rather than new, superprofit sectors. The South's boom was largely in textiles, shoes, autos, fish processing, lumber, soybeans, apparel, chemicals, and machinery—relatively mature sectors, many of which were declining nationally. The possibility that this tardy industrialization could be cut short by further migration overseas or heightened import competition resulted in pressure to keep labor costs, public sector taxes, and social wages low.

Out of this cauldron came a new regional organization, the Southern Growth Policies Board. Formed in 1971 by nine southern governors, from states that had once belonged to the Confederacy, the new board had a dual thrust. On the one hand, it tackled an unprecedented number of modern growth-management problems. On the other, it aspired to supplant intraregional competition with collaboration in traditional state economic development promotional activities. Initially, growth management concerns took priority. From its formation until 1976, the board's major proposals concerned coastal protection, environmental preservation, and water supply. In these years, the board was willing to accept restrictions on intensive development in return for the preservation of historical and environmental treasures (see Commission on the Future of the South, 1974; Godschalk et al., 1978).

By the mid-1970s, however, the Northeast-Midwest Congressional Coalition's attack on the "Sunbelt" and federal formulas grabbed the attention of the board, which opened a Washington office to fight the issues of interregional equity. For the next four years, it successfully fought for the retention of poverty and per capita income indicators in the formulas and blocked a "cost-of-living" weight that would have favored the Northeast (see Rafuse, 1977: 4; Southern Growth Policies Board, 1978; Weinstein, 1979).

A third shift in emphasis took place in 1980. Growth had slowed, gains in eliminating the national gap in per capita income had ended, and public officials were beginning to feel the incompatible pressures of overburdened public infrastructure and taxpayer revolt. An added problem loomed in the renewed unionization efforts in textiles and other manufacturing industries. With the economy largely dependent upon sectors that had been propelled from other regions by profit-squeezed imperatives, the region's leadership predictably worried about preserving the very characteristics that had induced growth in the first place. The final report of the energetic Commission on the Future of the South, in 1980, stated that southern planning efforts should respond to "a drive to solve problems at home" and that the time had come for "consolidation of growth and for looking inward." Interregional conflict was repudiated as a diversion, and the focus went back to conflicts at home.

Prominent among the commission's goals were methods for encouraging an inflow of capital to support growth, including foreign investment, cutting the fat out of business and political performance, and defeat of "all proposals to restrict the ability of industry to relocate, to impose restraints on the exercise of workers' rights, and to impose restrictive energy and environmental policies." The board had backed off from its previous concerns with the environment and had adopted an unabashed pro-economic-development stance (Southern Growth Policies Board, 1978: 10-12, 32-33).

The South in the 1980s, then, demonstrates the political tendencies of a region with hybrid growth. A previously underdeveloped region, it had boomed for two decades, but with relocating profit-squeezed industries, rather than new, innovative sectors. In the absence of institutions bred of class conflict—no union representation was considered necessary by the board in any of its activities—and building on a strong one-party system, the governors of the region found it easy and advantageous to get together to address internal problems of growth management. When attacked by the Northeast, they responded, and rather successfully. But in the longer run, their attentions were first and foremost dedicated to the problems of internal economic development. In the end, circumscribed by the fragile and cost-conscious nature of their growth sectors, they abandoned most of their growth-management initiatives for good old boosterism.

WHAT CONDITIONS INDUSTRIALLY INDUCED POLITICS?

The above discussion offers several hypotheses about protagonist factions, planning issues, and political targets of regional struggles associated with distinct economic structure. These operate at a level of generality that permit an elementary understanding of specific instances of regionalism. In reality, each historical case of regionalism possesses a high degree of specificity; in fact, each is unique. Some instances of regionalism may arise primarily from cultural or political impulses. It is not the aim of this chapter to explain all cases of contemporary regionalism. The aim is rather to explore the economic roots of regional politics in the postwar United States. In keeping with the commitment to regional complexity, this final section addresses how regional economic differences combine with social and political structures to heighten or dampen regionalist activities.

This synthesis is crucial, because similar economic events may evoke a wide range of responses across regions, from silence to full-scale regional war, depending upon the era, the region's history, its culture, and the political possibilities. In what follows, I review a set of additional economic, social, political, and cultural features of regions that interact with industrial restructuring to shape regional strife. The purpose is to show how these features form a filter between economic change and regional politics. In a similar manner, economic structure may act as a catalyst or a retardant to culturally or politically motivated cases of regionalism.

ADDITIONAL ECONOMIC FACTORS

In addition to sector composition and maturity, there are other aspects of economic change that are systematically related to regional politics. First among these is the pace of change experienced at any particular moment. If a region's economic growth or decline is stretched out over a long period of time and is relatively modest in its impacts, the provocation of regional politics is much less likely. It is where the rate of employment change is most sudden and intense that regional politics is most likely to erupt, as in the Youngstown case cited above. The geographical spread of change is also important in determining the intensity and scale of regional response. If employment growth or cutbacks are highly localized, regional politics will tend to operate on a local level. If employment cutbacks are widely spread, even if differentially, the pressure for a response at any one location will be less, but the possibilities for a broader regional or interregional coalition are greater.

Moreover, regional organizing around unique economic dilemmas may be transformed into interregional struggle if decline in one region occurs in tandem with growth in another. If the disparate growth paths are the product of a set of mature sectors moving from one to the other, it may become particularly acute. This is the crux of the postwar north-south antagonism. If the neighbors' growth take place in complementary sectors, the conflict may also be heated, especially if those sectors sell price-escalating inputs to a hard-pressed set of home industries. This is the crux of postwar east-west conflict.

The size of plants in a local or regional economy is an important ingredient in regional response. Those sectors with particularly large operations, in either the growth or the decline stage, are more apt to provoke place-based politics around entry or exit. An exodus of many small firms, such as the shoe industry from New England, is more difficult to respond to than the exit of one large corporation that closes out or carries with it many jobs. This is partly because of visibility, but also because of the unity of the work force that prevails as a fact of its common employer. Similarly, small coal mines in Appalachia, even if their environmental and resource-competing consequences are similar to those of large mines in the West, are less apt to fuel a regionwide movement. Small firms, even if entering or leaving a region in tandem, tend to call forth fragmented local responses by workers or community groups, rather than regionwide response.

The geographical locations of ownership of dominant sectors are also contributors to the shape of regional conflict. Economies with a high incidence of local ownership will produce more complicated local responses than sectors with absentee ownership. In the former case, local management and owners can be expected immediately to defend plant closings or new activities and to be powerful in local and regional political forums. They may, on the other hand, favor control of overly rapid development or maintenance of existing operations when an absentee ownership might not. Bluestone and Harrison's finding, that cases of corporate ownership showed lower rates of net job destruction than small businesses or conglomerates, seems to suggest this.[23]

CLASS STRUCTURE

Perhaps the most important key to the response of a community to a dominant change in its economy is its class structure. At its simplest, this means the division of the population among capitalist, small business (including farmers), and working classes. Each class has a direct material interest in community change; class factions may furthermore be separable along sectoral lines. Second, the question of class homogeneity is important. Business classes can be divided over issues other than those attached to a particular sector; if so, these cleavages may play an important role. A working class can be divided occupationally, by gender, by race, and by ethnicity. The greater these differences, the less likely that unity around a common regional position will occur. Instead, the negotiation of conflict will tend to take place within the region or within the sector with significant intraclass antagonism. Cases where plant-closing struggles devolve into fights among union members about the distribution of the severance package are examples. Any study of sectorally based regional politics must include an investigation of class structure.

Class structure also has a dynamic to it. As the sectoral composition of an economy changes, so does the proportional representation of groups within it. Long-term stable class structures are more apt to evoke a protest from abrupt economic change, whereas regions that have already been destabilized through massive in- or out-migration will be less apt to engender a self-conscious regional politics. Thus the framework for study includes an overview of changes in class structure over the last few decades and the importance of the sector in question to that change.

The identification of class composition does not necessary predict political response. Class location is not synonymous with class consciousness. Often, ethnicity, religion, culture, experience, or sexism intervene to temper or override the class interests that a simple model might predict. Again, the proof of the class concept lies in the case-by-case study of regional politics.[24]

SECTORAL SPECIALIZATION

In addition to its class composition, the degree of specialization in a community or region's industrial structure is a principal determinant of its susceptibility to regional politics. Simply put, the hypothesis here is that the more internally homogeneous the community's industrial structure, the more distinct it is from that of other regions; and the larger the share of the region's jobs that fall into such sectors, the greater the tendency toward a sectorally based regional politics.[25] Clearly, a steel plant closing in a small city will have a more disruptive effect than one in a large city. And two cities of equal size will experience a plant's closing differently depending upon the degree of diversification in the economy. Steel closings in Youngstown, Pittsburgh, and Chicago are examples. Youngstown is the smallest of the three, and the most highly dependent on steel. Pittsburgh is much larger, though its dependence on steel is quite great. The Chicago area is a large steel producer, but its economy is much more diversified than the other two. Steel plant closings have provoked greater concern and more identifiable movements in Youngstown and Pittsburgh than in Chicago.

The significance of local industrial structure is not merely a function of employment in those sectors whose dynamics are unsettling. It is also a function of the degree of linkages between those sectors and others in the economic base. Sectors that sell to and buy directly from the sector in question may be so tightly tied that their economic fortunes move in the same direction. Thus the total ensemble of industries tied together becomes the focus of an investigation of industrial structure. Regions may vary substantially in the degree of such intersectoral linkages, even around a particular industry. Some auto assembly plants are located near parts producers; others are not. Some steel mills are located near fabricators; others are not. Some coal mines spawn coal-fired electricity plants; others do not. The more spatially concentrated such interindustry links are, the more apt is spatial conflict to emerge over sectoral dynamics. Conversely, while growth or deterioration in one sector will affect the fortunes of a supplier or customer elsewhere, agitation around this linkage will not be registered in a territorial form.

EXISTING POLITICAL AND CULTURAL INFRASTRUCTURE

In addition to these strictly economic features of a region, institutional and organizational characteristics become important shapers of regional politics. These can be grouped roughly into four types: economic institutions (labor unions, business organizations, agricultural societies); political institutions (local and regional governments, political parties); cultural institutions (churches, community groups); and ideological institutions (the press, universities). Members of these organizations inevitably become involved in regional politics. Even if new organizations spring up, their membership tends to consist of building blocks of preexisting organizations. Understanding the nature, ideology, relative strength, and leadership potential of each of these is central to understanding the outcome of a specific regional struggle.

A brief catalog of these groups, with emphasis on their importance in different circumstances, is in order. Economic groups are clearly of paramount importance. It is these groups that mold internal class unity or dissension on strategy. They are also the basis of representation on many types of economic development bodies that may play a central role in the unfolding of regional politics. Chief among them are labor unions. In addition to the relative strength of unions among the working-class population in an area, the degree of unity and organization across the region's different unions is an important factor (e.g., the presence of a strong state or regional federation) and the degree of unity or dissension between union locals and their internationals. Business groups are also pivotal institutions.[26] Chief among these is the traditional chamber of commerce, which tends to represent the commercial/financial/real estate classes. In cases of rapid regional change, other business groups may present a significant challenge to the former group; farmers' or ranchers' organizations in energy regions, and manufacturers' organizations, are examples.

Political institutions of paramount importance are the jurisdictional units that govern the area. If a disrupted region is coterminant with one or more local governments, then its chances of pursuing a solution to regional conflict at that level, or at least gaining the support of local government, are enhanced. The powers of that wing of the state are relevant as well—what powers to tax, spend, adjudicate, and regulate land use does it have? The extent of local bureaucracy, its structure, and the quality of personnel, including locally elected officials, also shape regional politics. Frequently the pressure to create new regional governments or special services districts, or to create state or federal agencies with powers to intervene in land-use and infrastructure decisions, comes from the misfit between existing political structure and the nature of the sectorally induced problem.

Political parties at the local and regional levels also form a critical element in regional sensitivity. In some regions, one party may long have dominated politics; in this case, its refusal to espouse a regional issue will be very difficult to counter. But if it does take up the regional banner, it provides a preexisting structure for political cohesion. This occurred in the case of the dominant party, the Democrats, in the formation of the Southern Growth Policies Board in the 1970s. In other regions, strong and competitive partisan politics may polarize attempts to achieve internal regional unity. This occurred in the Northeast in the 1970s and explains in part why the region's congressional caucus, rather than the governors, provided the leadership for the coalition. In yet other regions, party structure may be competitive yet weak, creating many openings for new regional coalitions. This was the case in the intermountain West in the 1970s, enabling governors from both major parties to join forces, when their counterparts in the Northeast could not.

Cultural organizations, for want of a better umbrella designation, include community groups of diverse types, environmental groups, and religious organizations. In many cases of regional politics, such groups have been important foci of organizing efforts. Cultural organizations such as churches and community groups can be vehicles for cross-class coalitions that are difficult to engineer via purely economic organizations. This was particularly true in the Youngstown and Pittsburgh plant closings movements. Cultural contributions or constraints or the formation of regional politics operate through more informal channels as well—folklore, belief systems, attitudes toward authority, dialects or language differences, and so on.

Finally, the role of those institutions whose product is knowledge, information, and/or ideology in a regional political coalition may be extensive. Newspapers, which are private enterprises with both a probusiness and a prolocal ideology, are generally the major opinion creators in the region as a whole. Their commitment to cover and their editorial position on attempts at regional organizing can have a large impact. Universities and other nonprofit research institutions may lend expertise to one or another regional group in its effort to muster support and documentation for its position.

CONCLUSION

Regional conflicts have frequently dominated the American political scene. In the United States, these have predominantly been economically induced, where regional growth rates diverge due to underlying differences in the dynamics of their industrial structures. New regionwide organizations have emerged to cope with problems associated with industrial restructuring, especially in areas where uneven development has been particularly intense. In this chapter, I have argued that the distinguishing characteristics of such industrially prompted regional politics vary according to whether a region is experiencing industrial decline or expansion, and, if the latter, whether the growth is based on innovation or on relocation.

For some, the analysis presented here may appear overly schematic and economically deterministic. I have, therefore, further probed the conditions under which stressful industrial restructuring might be subdued or exacerbated, given the ownership, size, and dominance characteristics of industries at issue and the political and cultural milieu within which such events are embedded. For the United States in the twentieth century, the industrial restructuring model appears to work fairly well, especially for the combative 1970s.

NOTES

1. This chapter draws heavily upon Markusen (1987b: chaps. 5, 7, 8, and 9).

2. This section summarizes the argument presented in Markusen (1985). The analysis here draws upon the Marxist, Schumpeterian, and product cycle traditions, which are discussed at length in Markusen (1986).

3. See Mandel (1975) for an extended discussion of superprofits.

4. Taylor (1975) argues that most firms originate in the founder's hometown, although Feller (1974) finds that there is no necessary relationship between inventive activity and subsequent location.

5. On the relationship between innovation and agglomeration, see Thompson (1965), Friedmann (1973), and Pred (1976).

6. For an elaboration of this change in strategy, see the product cycle literature, originating with Kuznets (1930) and Burns (1934) and reformulated by Vernon (1966) and Hirsch (1967).

7. The normalcy of agglomeration followed by dispersion was first hypothesized by Hoover (1948). See Persky (1978) and Hansen (1980) for a debate over the extent to which postwar north-south differentials can be attributed to product life cycle spatial filtering. Bergman and Goldstein (1983) show that noncyclical structural change has been a prominent feature in U.S. metropolitan shifts in the 1970s.

8. Another important stage, that of oligopolistic profits, is not included in this discussion because its role in regional politics is not discussed in the present chapter.

9. See the separate accounts of the textile and shoe industries in Markusen (1985: chaps. 9 and 10). Harrison (1984) traces the out-migration of these sectors from New England.

10. See Saxenian (1980) for empirical documentation of this phenomenon in Silicon Valley.

11. Lynd (1982) emphasizes the importance of a tradition of rank-and-file dissidence as a factor in Rustbelt organizing. Metzgar (1980) stresses the same factor in explaining the difference between the community response to the Youngstown and the Johnstown steel closings in the 1970s. Simon (1980) emphasizes ideology as a factor in the failure of a worker-community-ownership strategy to win adherents in Appalachia.

12. Masterful works combining the cultural with the economic have been produced by some of those working on Wales (Cooke, 1981) and, more recently, for Appalachia (Clavel, 1982; Whisnant, 1979).

13. For an extended discussion of how regionalism is shaped by American political structure, see Markusen (1986: chap. 3).

14. These two powers of the state are universal across the United States. In addition, the state may become involved in other ways, such as through economic development programs with special subsidies, through urban-renewal-type programs, or through labor training.

15. See Schoenberger and Glasmeier (1980) for an account of the MX conflict.

16. For accounts of these responses, see Gold (1974), Gottlieb and Wiley (1982), Massey (1980), Moen et al. (1981), and Toole (1976).

17. Western Governors' Policy Office (1978: 2, 26-27, 39), Polsky (1979: 67). See also the extended analysis in Markusen (1987b: chap. 9).

18. The Kelly and Shutes (1977) analysis of Lyke's role in Youngstown, for instance, led to an overly anticorporatist interpretation, as if another, nonconglomerate owner would have acted differently. The subsequent closings of U.S. Steel in the same community and in its hometown of Pittsburgh demonstrate the limits of blaming conglomeratism, and shows that the problem may be endemic to steel under its capitalist form of organization.

19. For a provocative view of how product cycle maturation has produced political pressures at a national scale, see Kurth (1979). Because the appropriate territorial scale is increasingly worldwide, there is some question as to the effectiveness of even national planning in a world capitalist system. Harvey (1982: chap. 13) argues that the system as a whole is headed for global crisis and increasing geopolitical instability.

20. For an overview of the New England experience, see Harrison (1984). Markusen (1987a) covers the case of the Midwest in the 1980s.

21. See Peirce (1976) and Rafuse (1977) on the origins of the northeastern coalition.

22. On the evolution of the southern economy, see Danhoff (1964) and Naylor and Clotfelter (1975).

23. Bluestone and Harrison (1982). One can argue, on the other hand, that conglomerates consistently buy up the already mature plant with the express purpose of extracting liquid capital by running it into the ground. If this is the case, then the problem is not just the conglomerate form, but the underlying dynamics of that sector in that location.

24. See Plotke (1980) for an extended discussion of this point applied to the United States.

25. A corollary to this point is that community decision making is more pluralistic when its economy is more diversified, an argument made by Prethus (1964) and Clark (1974). Young and Newton (1980) illustrate the nondiversified case with five single-industry case studies.

26. See Arsen (1981) for an overview of the literature on business behavior in local government politics.

REFERENCES

ARSEN, D. (1981) Public Choice and Business: The Impact of Firms on Municipal Budgets. Ph.D. dissertation, University of California, Berkeley.

BALDWIN, M. (1973) The Southwest Energy Complex: A Policy Evaluation. Washington, DC: The Conservation Foundation.

BERGMAN, E. and H. GOLDSTEIN (1983) "Dynamics and structural change in metropolitan economies." American Planning Association Journal (Summer): 263-279.

BLUESTONE, B. and B. HARRISON (1982) The Deindustrialization of America. New York: Basic Books.

BURNS, A. F. (1934) Production Trends in the United States. New York: Bureau of Economic Research.

CLARK, T. (1974) "Can you cut a budget pie?" Policy and Politics (2): 3-32.

CLAVEL, P. (1982) Opposition Planning in Appalachia and Wales. Philadelphia: Temple University Press.

Commission on the Future of the South (1974) The Future of the South. Chapel Hill, NC: Southern Growth Policies Board.

COOKE, P. (1981) "Class relations and uneven development in Wales." Working Paper. Cardiff, UK: University of Wales, Institute of Science and Technology, Department of Town Planning.

DANHOFF, C. (1964) "Four decades of thought on the South's economic problems," pp. 7-68 in M. Greenhut and W. Tate Whitman (eds.) Essays in Southern Economic Development. Chapel Hill: University of North Carolina Press.

FELLER, I. (1974) "The diffusion and location of technological change in the American cotton-textile industry, 1890-1970." Technology and Culture 15 (4): 569-593.

FRIEDMANN, J. (1973) Urbanization, Planning and National Development. Beverly Hills, CA: Sage.

GODSCHALK, D., B. KNOPF, and S. WEISSMAN (1978) Guiding Growth in the South. Research Triangle Park, NC: Southern Growth Policies Board.

GOLD, R. (1974) A Comparative Study of the Impact of Coal Development on the Way of Life of People in the Coal Areas of Eastern Montana and Northeastern Wyoming. Missoula, MT: Institute for Social Science Research.

GOTTLIEB, B. and P. WILEY (1982) Empires in the Sun: The Rise of the New American West. New York: Putnam.

HANSEN, N. (1980) "Dualism, capital-labor ratios and the regions of the U.S.: a comment." Journal of Regional Science 20 (3): 401-403.

HARRISON, B. (1984) "Regional restructuring and 'good business climate': the economic transformation of New England since World War II," in L. Sawers and W. K. Tabb, (eds.) Sunbelt/Snowbelt: Urban Development and Regional Restructuring. New York: Oxford University Press.

HARVEY, D. (1982) The Limits to Capital. Oxford: Basil Blackwell.

HAVEMANN, J., R. STANFIELD, and N. PIERCE (1976) "Federal spending: the North's loss is the Sunbelt's gain." National Journal (June 26): 878-891.

HAYES, L. (1980) Energy, Economic Growth and Regionalism in the West. Albuquerque: University of New Mexico Press.

HIRSCH, S. (1967) Location of Industry and International Competitiveness. London: Oxford University Press.

HOOVER, E. M. (1948) The Location of Economic Activity. New York: McGraw-Hill.

"Jobs and revenues shift toward Sunbelt." (1976) New York Times (April 7).

KELLY, E. and M. SHUTES (1977) "Lykes' responsibility for closing the Youngstown Campbell works." Cleveland: Ohio Public Interest Campaign (November).

KURTH, J. (1979) "The political consequences of the product cycle." International Organization 33 (1, Winter): 1-34.

KUZNETS, S. (1930) Secular Movements in Production and Prices. Boston: Houghton Mifflin.

LURIA, D. and J. RUSSELL (1981) Rational Reindustrialization. Detroit: Widgetripper Press.

LYND, S. (1982) The Fight Against Shutdowns: Youngstown's Steel Mill Closings. San Pedro: Singlejack Books.

MANDEL, E. (1975) Late Capitalism. London: New Left Books. (Originally published in French, 1972.)

MARKUSEN, A. (1985) Profit Cycles, Oligopoly and Regional Development. Cambridge: MIT Press.

MARKUSEN, A. (1986) "Empirical research in the Marxist and Schumpeterian traditions: reflections on explaining spatial change," in S. Helburn and D. Bramhall (eds.) Marx, Schumpeter, Keynes: A Centenary Celebration of Dissent. New York: M. E. Sharpe.

MARKUSEN, A. (1987a) "Regional disparities in the 1980s: the roles of trade and cold war." Working Paper. Evanston, IL: Northwestern University, Center for Urban Affairs and Policy Research.

MARKUSEN, A. (1987b) Regions: The Economics and Politics of Territory. Totowa, NJ: Rowman & Littlefield.

MASSEY, G. (1980) "Critical dimensions in urban life: energy extraction and community collapse in Wyoming." Urban Life 9 (2): 187-199.

METZGAR, J. (1980) "Plant shutdowns and worker response: the case of Johnstown, Pennsylvania." Socialist Review 53: 9-49.

MOEN, E. et al. (1981) Women and The Social Costs of Economic Development: Two Colorado Studies. Boulder, CO: Westview Press.

NAYLOR, T. and J. CLOTFELTER (1975) Strategies for Change in the South. Chapel Hill: University of North Carolina Press.

Northeast-Midwest Institute (1978a) Focus on Regional Cooperation: North-South Summit, Boston, August 24-25. Washington, DC: Author.

Northeast-Midwest Institute (1978b) Per Capita Income Is Not a True Measure of Health. Washington, DC: Author.

Northeast-Midwest Institute (1980) The Effects of Rising State Severance Tax Revenues: 1980-1990. Regional Energy Impact Brief No. 10. Washington, DC: NMI, House Annex.

PEIRCE, N. (1976) "Northeast governors map battle plan for fight over federal funds flow." National Journal (November 27): 1695-1703.

PEIRCE, N. and J. HAGSTROM (1977) "Western states join in forming a united energy front." National Journal (February 5): 208-210.

PERSKY, J. (1978) "Dualism, capital-labor ratios and the regions of the U.S." Journal of Regional Science 18 (3): 373-381.

PLOTKE, D. (1980) "The United States in transition: toward a new order." Socialist Review 54 (November-December): 71-124.

POLSKY, A. (1979) "The American West looks east." Far Eastern Economic Review (November 2): 56.

PRED, A. (1976) "The interurban transmission of growth in advanced economies: empirical findings vs. regional-planning assumptions." Regional Studies 10: 151-171.

PRETHUS, R. (1964) Men at the Top: A Study in Community Power. New York: Oxford University Press.

RAFUSE, R., Jr. (1977) The New Regional Debate: A National Overview. Washington, DC: National Governors' Conference.

SAXENIAN, A. (1980) Silicon Chips and Spatial Structure: The Semiconductor Industry and Urbanization in Santa Clara County, California. M.A. thesis, University of California, Berkeley, Department of City and Regional Planning.

SCHOENBERGER, E. and A. K. GLASMEIER (1980) "Selling the MX: The air force asks Nevada to move over." Progressive (May): 16-21.

"A second war between the states." (1976) Business Week (May 17): 92-113.

SIMON, R. (1980) "The labor process and uneven development: the Appalachian coalfields." International Journal of Urban and Regional Research 4 (1, March): 46-71.

Southern Growth Policies Board (1978) Annual Report. Chapel Hill, NC: Author.

TAYLOR, M. J. (1975) "Organizational growth, spatial interaction and locational decision-making." Regional Studies 9: 313-323.

THOMPSON, W. (1965) A Preface to Urban Economics. Baltimore: Johns Hopkins University.

TOOLE, K. R. (1976) The Rape of the Great Plains. Boston: Little, Brown.

Tri-State Conference on Steel (1984) A Community to Save Pittsburgh's Steel Industry. Homestead, PA: Author.

VERNON, R. (1966) "International investment and international trade in the product cycle." Quarterly Journal of Economics, pp. 190-207.

WEINSTEIN, B. (1979a) "Cost-of-living adjustments for federal grants-in-aid: a negative view." Chapel Hill, NC: Southern Growth Policies Board. (February)

WEINSTEIN, B. (1979b) "The current recession and its impact on the South." Testimony before the Subcommittee on Fiscal and Intergovernmental Policy of the Joint Economic Committee, U.S. Congress. (mimeo, October 16)

Western Governors' Policy Office (1978) Balanced Growth and Economic Development: A Western White Paper, Vol. 1. Denver, Colorado: WESTPO.

WHISNANT, D. (1979) Modernizing the Mountaineer: People, Power and Planning in Appalachia. New York: B. Franklin.

YOUNG, J. and J. NEWTON (1980) Capitalism and Human Obsolescence. Montclair, NJ: Universe Books/Allanheld, Osmun and Co.

5

Local Politics and the Employment Relation: Construction Jobs in Philadelphia

ROBERT A. BEAUREGARD

IN 1985, a Philadelphia city council member introduced a bill that would set aside construction jobs for residents, minorities, and women. The bill was instigated by a coalition of human rights organizations, neighborhood associations, employment advocates, feminists, gay/lesbian groups, labor unions, and churches. In Boston, Chicago, and other large cities, similar groups had emerged to champion such hiring preference legislation. Their goal was to pressure local government to intervene at the workplace in order to reshape the employment relation between worker and owner, thus obtaining private sector job commitments. In effect, city governments would be attempting to influence the labor process.

Certainly, this was a new political demand, and a new role, for local governments. The excesses of industrial capitalism in the 1800s and early 1900s had caused the state to intervene in the labor market. Laws pertaining to hours of work, overtime, child labor and female work, and workmen's compensation had been passed, later to be joined by minimum wage legislation, unemployment insurance, collective bargaining protection, health and safety regulations, and equal employment laws. At no time, however, has the state seriously challenged the notion that "firms own the jobs" and "have the unilateral right to hire and fire at will" (Clark et al., 1986: 236-240).

AUTHOR'S NOTE: *I would like to thank Mickey Lauria for his very helpful comments on an earlier draft of this chapter.*

Until recently, state intervention in labor relations emanated primarily from the federal and state levels. At the local level, government had neither the power nor the legitimacy, much less the inclination, to become so involved. Large corporations can bring political influence to bear on state and national politicians thus easily bypassing local governmental authority, while both small and large business establishments can often relocate to other jurisdictions. Equally as important, the need to maintain an integrated national economy and a political nation have engendered constitutional limitations on interstate commerce as well as the constitutional protection of spatially unbounded "rights" of individuals, thus placing severe constraints on any local governmental attempts to regulate commerce or employment (Clark, 1981a).

With the passage of equal employment opportunity legislation in the 1960s and early 1970s, and the growth of government employment, local government involvement in workplace concerns increased. Although equal employment laws were promulgated nationally, local governments established human rights commissions to oversee and monitor their application. At the same time, job training and public service employment turned city and county governments into significant actors within the low-wage sector of the local labor market. Simultaneously, the emergence of the government as a major employer, particularly in the larger cities, brought it into direct confrontation with newly emerging "white-collar" labor unions. Though not a capitalist, the local government still found itself functioning as management.

The restructuring of local economies throughout the postwar period, made apparent by the recession of 1973-1975, further highlighted the burgeoning association of local governments with workplace concerns. The movement of the economic base away from manufacturing and toward service jobs, the concomitant polarization of the income distribution, the shrinkage of overall labor demand in many older cities, and the recomposition of the resident labor force positioned the local government as a potential actor in the local labor market. Grass-roots political activity focused on the local government. For example, blacks expected preferential treatment from black mayors in municipal hiring, and demanded affirmative action on the part of construction unions. In effect, a variety of forces, with economic restructuring and the spatial recomposition of the urban labor force playing key roles, provided the opportunity for the local government to become a participant in the labor process. Simultaneously, those same forces made it necessary for certain segments of the urban labor force (particularly minorities) to turn to the local government for assistance in their quest for decent employment.

The aim of this chapter is to explore how grass-roots political activity in a single city, Philadelphia, came to be focused on workplace issues and yet was channeled through the mediating agency of the local government rather than directly targeted on employers themselves. The overall intent is to (1) illuminate the political consequences and social concomitants of economic restructuring, (2) reflect upon the current nature of grass-roots political action as it confronts the prerogatives of capitalism and leans heavily upon legal constraints and racial imperatives, and (3) demonstrate the importance of the spatiality of urban development in these processes. This will be achieved by focusing on hiring preference ordinances as they emerged in the 1980s, specifically as those ordinances responded to racial discrimination.

The chapter begins with a description of the emergence of hiring preference ordinances as local governmental policy, describing in detail the nature of the bill in Philadelphia and the political action that accompanied it. We then turn to an analysis of the forces whose conjuncture created the possibility for collective political action and focused such activity on the local government. Particular attention is given to the racial and legal dimensions of these events and to the contextual influence of economic restructuring on both the demand side of the labor market and on the city's resident labor force. This understanding is then used to reflect upon the current interest in urban social movements, collective consumption, and economic restructuring. The politics of hiring preference ordinances is, from the dominant perspective, somewhat of an anomaly.[1] It concerns labor relations, but organizing and political events take place outside the workplace. The state is the target of political mobilization, but the issue is not simply one of reproduction, or collective consumption. Moreover, the local government is taking the initiative in a policy arena heretofore the prerogative of the national government, and not only as a mediator but also as a market participant. As a singular event, then, political mobilization around hiring preference ordinances presents a theoretical challenge to current thinking in urban political economy.

THE BACKGROUND

State governments and the federal government have been involved for decades in the employment relation, that is, the contractual relationship between labor and capital specifying the conditions of employment (Clark, 1981b: 413, 416-418; Steinberg, 1982: 3-13, 60, 109-118). Prior to the 1930s, legislation dealt mainly with standards for wages and hours of work. Most were state laws. The federal government did not become involved in the employment relation until the New Deal. The National Labor Relations Act (1935) established the rights of workers to form unions and engage in collective bargaining, and the Fair Labor Standards Act (1938) instituted a national minimum wage and flexible hours standards. As Steinberg (1982: 8) has written: "Wage and hour standards laws constitute . . . explicit government intervention—at least formally—on behalf of employees into the setting of the terms and conditions of the labor contract." Such intervention, of course, is incompatible with a strict reading of the freedom-of-contract doctrine, which argues that the government should not interfere with, and in fact should protect, the right of individual employers and employees to unrestrained negotiation over the conditions of employment. That is, the doctrine assumes that each participant enters willingly into the employment contract, both having struck an acceptable bargain.[2]

Equal employment legislation further eroded the freedom-of-contract doctrine. It regulates the grounds on which an employer can reject an applicant as unsuitable for hiring, while the legislation's affirmative action component constrains who is ultimately hired. Beginning in the late 1930s, the Supreme Court became increasingly sympathetic to plaintiffs claiming employment discrimination (Burstein, 1985: 13-18), thus setting the context for President Roosevelt's executive order no. 8802 in 1941, which prohibited racial discrimination by the federal government and by defense contractors.[3] In 1945, New York State adopted a law prohibiting employment discrimination on the basis of race, religion, and national origin. Still, three decades went by before a national equal employment opportunity law was passed. The Civil Rights Movement of the 1950s and 1960s, changing public attitudes, and numerous instances of racial violence eventually created a groundswell of support for national equal employment opportunity legislation (Burstein, 1985: 13-96).

Title VII of the Civil Rights Act of 1964 was the first comprehensive response to employment discrimination (Citizens' Commission on Civil Rights, 1984; Fusfeld and Bates, 1984: 122-124; Levitan et al., 1975: 267-275). It extended coverage beyond the federal government and defense contractors to private employers who are not government contractors and to labor unions and employment agencies as well, and prohibited discrimination on the basis of race, religion, national origin, and sex. Moreover, the Civil Rights Act created the Equal Employment Opportunity Commission (EEOC) to enforce Title VII. Eight years later, the Equal Opportunity Employment Act strengthened the enforcement provisions of Title VII by defining more clearly the types of actions that could and could not be taken by the EEOC. The EEOC can now take alleged discriminators to court, individuals are allowed to sue, and class action suits are permissible. Simultaneous with the emergence of national equal employment legislation, the Kennedy and Johnson administrations were issuing executive orders no. 10925 and no. 11246, respectively, to assure federal contractor compliance with equal employment laws through the establishment of the Office of Federal Contract Compliance (OFCC). By the 1980s, the federal government had established equal employment opportunity legislation as another mechanism for intervening in the employment relation. The intent of these laws was to foster nondiscriminatory hiring practices and, through goals and affirmative action plans, assure representation of various groups in the ranks of the employed.[4]

CONSTRUCTION INDUSTRY

One industry that received special attention during the 1960s and thereafter was the construction industry. Ray Marshall, Secretary of Labor under President Carter, wrote in the late 1970s that "in the past decade, the construction industry has received more nationwide attention with regard to racial employment than perhaps any other sector of the economy" (Marshall et al., 1978: 26). Construction trades and crafts have traditionally been guildlike in their activities, exerting strong control over entry and discriminating against minorities and women in terms of union membership, job referral, and training and apprenticeship opportunities. "Traditions of nepotism and overt racial discrimination among construction unions, coupled with exclusive bargaining and referral arrangements these unions had with the major construction contractors, virtually excluded minorities from employment on government construction projects" (Citizens' Commission on Civil Rights, 1984: 41). President Kennedy in 1963 had ordered a review of black employment on federal construction projects. The results reinforced his commitment to civil rights legislation but did not lead to specific action (Chafe, 1986: 215). In 1965, the national government targeted the construction industry for special attention and the OFCC established four "special area programs" in St. Louis, San Francisco, Cleveland, and Philadelphia in order to assure compliance through affirmative action measures. Local minority groups in these and other cities (e.g., Cincinnati, Washington, DC, Pittsburgh) confronted the construction unions and demanded changes in hiring practices, apprenticeship and other training programs, and the institution of employment goals and timetables (Dubinsky, 1973; Marshall et al., 1978: 26-61).

Under either federally imposed plans or "hometown" plans, contractors and subcontractors were required to prepare manning tables indicating the number of jobs, by trade and craft, that would be (in the preaward phase) or were being (in the construction phase) given to minorities. Construction unions were encouraged to set up training programs for minorities, including taking on more minority apprentices, to change hiring practices in order to provide equal opportunity to all, and to remove barriers preventing minorities from moving from apprenticeship to journeyman status. Such changes were to apply not just to the skilled crafts but also to laborers as well.

The desirability of construction employment for minorities was unquestioned. Union status assured workers of relatively high wages, even though work was not just seasonal but often erratic even during the optimum months for construction. Access to training and apprenticeship programs conferred skills (e.g., carpentry) that could not only be used on construction projects but might also be transferred to other settings (e.g., a cabinetmaking shop). Even employment as a laborer meant higher wage rates than could be obtained in most other industries, particularly with a declining manufacturing base in many cities.[5] Moreover, being a laborer does not demand education beyond high school or any special skills, and a number of the trades (e.g., bricklayers, roofers, and lathers) require only a minimum of formal education. Here is a category of employment that could easily absorb minority workers. Equally important, however, the construction industry then represented (and still does) a bastion of discrimination. Minorities and women were clearly unwelcomed. Ending discrimination here was an important strategic and symbolic decision.

HIRING PREFERENCE ORDINANCES

Hiring preference ordinances are an extension of this struggle for equal employment opportunity in the construction industry, with some significant new overtones. The various affirmative action plans instituted in cities across the country had had little impact upon the access of minorities and women to construction employment (Levitan et al., 1975: 272; Marshall et al., 1978: 4). In Philadelphia, race-conscious hiring goals were initiated in a 1969 plan and were expressed in terms of "good faith efforts" on the part of contractors and subcontractors. After eight years of operation, however, seven construction trades in the metropolitan area still required special procedures to bring them into conformance.[6] Subsequently, new regulations and laws continued to strengthen equal employment procedures, but with no significant shift in the hiring patterns within Philadelphia's construction industry (Citizens' Commission on Civil Rights, 1984: 45-65).

In 1986, a retrospective look at the Philadelphia Plan by *The Philadelphia Inquirer* found that minorities and women were still not represented on construction projects in ratios compatible with their availability in the labor force and their presence in the resident population (Collins, 1986). Construction firms holding federal contracts in excess of $10,000 were to have allocated 17.3% of the hours in each craft to minorities and 6.9% of the hours to women. Yet, a 1984 survey discovered that only minority laborers and cement masons exceeded the federal goal, with minorities in most crafts receiving between 3% and 16% of the hours. Women received no more than 3% of the hours in any craft. In comparison, blacks constituted 29.1% of all construction workers in the city in 1980 and Hispanics 2.4% for a minority total of 31.5%, and women 7.0% (Beauregard, 1987: 127). Progress was being made, but at a glacial pace.

Such findings were not unique to Philadelphia. Affirmative action plans in many other large cities had had similar results (Marshall et al., 1978). Systematic discrimination in the construction industry was deeply rooted, and withstood almost constant pressure from civil rights groups and the federal government throughout the 1960s and 1970s. In the late 1970s and early 1980s, the quest for equal employment opportunities in construction took another turn—hiring preference ordinances (see Beauregard, 1987). The leader was Boston (Clark, 1985: 85-110).

The Boston Plan had been established in 1969 to bring together unions, contractors, and minority organizations in a partnership that would offer training and apprenticeship programs for minorities, thus enabling them to enter into unionized construction work. The impetus was a coalition of black construction workers, United Community Construction Workers, who rallied around the slogan "Community Jobs for Community Workers." Success eluded them. Boston's construction industry was generally moribund in the 1970s and, without growth, even white construction workers were suffering from high unemployment rates. Still, in 1975, the city funded the Third World Jobs Clearinghouse (TWJC), an interracial community referral agency established by minority groups, to push harder for affirmative action in construction. The unions and contractors gave TWJC little support, and pressured the city government to defund TWJC. Only the intervention of the Justice Department prevented that from happening.

Founded in 1977, the Boston Jobs Coalition operated under the slogan "Boston Jobs for Boston People," thus shifting the emphasis from minority discrimination to the link between job development activities within the city and the employment needs of city residents. Political activity eventually resulted in Mayor Kevin White issuing an executive order which required that a preference in construction hiring, specifically on construction projects in which the city or the federal government was providing subsidies, be given to residents, minorities, and women. Thus the equal opportunity employment thrust, characterized as it has traditionally been by a concern with race and gender, took on a new spatial dimension: resident preference.

Local construction unions and contractors sued the city, contending that the executive order violated the Commerce Clause by making the city a regulator of the regional labor market.[7] This supposedly prevented the unrestricted movement of labor across jurisdictional boundaries and thereby weakened the national integration of the economy. Eventually the case landed in the U.S. Supreme Court, where the city successfully argued that it was not regulating the market but simply asserting its prerogatives as a market participant, thus falling back on decades of constitutional support for the rights of employers to maintain dominant control over the employment relation. The Boston executive order became a city ordinance in 1983, and subsequently has been extended to cover projects wholly funded through the private sector.

By the time the hiring preference issue had become politically salient in Philadelphia, other cities had also developed such ordinances. Chicago, for example, had passed its Public Works Hiring Order of 1984. Camden, New Jersey, legislated a hiring preference ordinance in 1980, only to have it challenged in court by the construction unions. An out-of-court settlement left the ordinance basically intact but added a training option for contractors and unions not able to achieve the targeted percentages.[8] A number of other cities also considered or passed hiring preference ordinances during this period.

As occurred elsewhere, the impetus for the Philadelphia bill, Bill 649, came mainly from the minority community with significant support from the city's Human Relations Commission.[9] In Councilman David Cohen, one of the most liberal members of the city council, advocates found an elected official willing to write and introduce such a bill, which he did in June of 1985. The bill noted the job loss suffered in Philadelphia since 1970 and the "egregious discrimination" by the local construction industry against minorities and women. Through the bill, the city hoped to promote the economic well-being of its citizens, relieve unemployment, ensure that public funds were used to benefit the city, combat discrimination, and provide equal employment opportunities for all persons. Thus any construction project funded in whole or in part with city funds or funds administrated by the city that exceeded $1,000,000 and any subcontract of a similar nature exceeding $25,000 had to comply with minimum employment preference goals of 45%-50% of the hours for city residents, 25%-35% of the hours for minorities, and 5%-8% of the hours for women. These goals, moreover, were to apply to all trades and crafts.

The bill's supporters expanded their base of operation through the establishment of the Bill 649 Coalition. In the ensuing months after introduction of the bill, the coalition lobbied the city council, held public rallies, and presented testimony. Throughout the remainder of the year, a good deal of media attention, particularly by the newspapers, was given to the hiring preference bill. Supporters included minority organizations, churches, gay/lesbian groups, feminist groups, labor unions, and the aforementioned Human Relations Commission. Few members of city council proclaimed their commitment. Two of the most powerful members of council, both black, avoided the issue, one introducing a similar but weaker bill and the other, the council president, remaining neutral.

Mayor W. Wilson Goode opposed Bill 649 as did the business community, most of the construction trade and craft unions, and the building contractors. The mayor argued that the bill would produce two undesirable consequences. First, it would encourage surrounding municipalities to pass similar ordinances, thus depleting the employment opportunities for Philadelphia's construction workers. Second, it would erode the positive business climate in the city by imposing undue regulations upon one particular industry. This second reason was basically the argument presented by the chamber of commerce. In representing business interests, it also claimed that the expansion of governmental powers would be detrimental to investment and only redound to the city's disadvantage. Last, there was the opposition of contractors and unions. The contractors were concerned about interference in hiring practices and thus loss of control over the manning of construction projects. In addition, they pointed out the difficulty of reaching the goals with so few minorities and women trained in the various trades and crafts. While the mayor was concerned with the resident goal and contractors with minorities and women, the primarily white, trade and craft unions opposed the minority goals. Councilman Cohen had specifically included the resident goal in order to assuage the unions, whose membership was drawn almost exclusively from the white "river wards" of the city. The tactic failed, and union opposition was pronounced.

The opposition of the mayor, the business community, and many unions convinced council members to withhold support from Bill 649. The hiring preference ordinance thus lacked the votes to be read out of committee. Despite citywide elections in November of 1987, and the potential they had for raising issues, the hiring preference ordinance failed to make an appearance. By early 1988, Bill 649 was still not law, and the Bill 649 Coalition was silent.

The second term of W. Wilson Goode could well bring passage of the hiring preference bill. Philadelphia is not that different from Boston, Chicago, and Camden in terms of the economic and political forces that gave rise to the issue and set the stage for local governmental involvement. Still, the above description leaves us with only a superficial understanding. If we are to position hiring preference ordinances within the current restructuring of American society, we must look more closely at how the issue was given objective definition and how the link between local politics and the employment relation was forged.

ANALYSIS

Why did the hiring preference ordinance become a political demand of grass-roots organizations in numerous cities across the country, specifically Philadelphia? Because it involves the labor process, we would expect that part of the answer lies with the economic restructuring that large cities experienced in the postwar period. Because it involves the state, we must also consider the policy transformations experienced by the state at the local and federal levels. Because minorities and urban residents are involved, we must not overlook how the population of cities has changed. These are the major lines of the inquiry. The analysis proceeds in terms of four specific questions: (1) Why did minorities direct their energies toward the local government? (2) Why did they target the construction industry? (3) Why was the local government willing to become involved? (4) To what extent is this a particularly "urban" issue? The answer to these questions will enable us to reflect theoretically on the nature of contemporary, urban political activity.

INVOLVING THE LOCAL GOVERNMENT

Across the postwar period, Philadelphia has experienced a loss of employment. From a postwar peak in 1953 to 1985, nongovernmental employment within the city fell by 23.2%, a net loss of 184,338 jobs.[10] On the other hand, the number of employed black residents increased from 171,594 to 201,377 between 1950 and 1980. This 17.4% increment to black employment, however, fell far short of the expansion of the black civilian labor force and the black population: 56.5% and 68.6%, respectively, over these four decades.[11] In effect, then, the rising supply of black workers in Philadelphia was faced with a shrinking job market. Moreover, when the expansion of black employment is compared to the growth of the black labor force, the employment prospects for blacks seem even dimmer.

While nongovernmental employment within the city was declining, however, government employment was on the upswing. Between 1960 and 1980, the number of city residents employed by local, state, and federal governments grew by 8.4%. The corresponding increase for black residents was 34.0%, thus enabling blacks to expand their share of governmental jobs from 30.4% to 37.6%. On the other hand, local government employment fell by 4,703 full-time equivalents (12.7%) from 1974 to 1983.[12] Despite that decline, the minority percentage of city workers has remained stable since the early 1980s (Beauregard, 1988). At the same time, the numbers of blacks employed in other industrial sectors—services; transportation, communications, and public utilities; and finance, insurance, and real estate—was also expanding. (See Table 5.1.) The growth in employment in government was particularly important for the turn to local government by grass-roots organizations concerned with affirmative action.

Throughout the decade of the 1960s and until approximately 1981, the federal government had been deeply involved in employment and training endeavors, most of which were carried out by local agencies (Claque and Kramer, 1976; Clark and Gertler, 1983). The Manpower Development Training Act of 1962 and the Comprehensive Employment and Training Act (CETA) of 1973 established job training programs targeted at the unemployed and economically disadvantaged in num-

TABLE 5.1
Industry of Black Employed Residents of Philadelphia, 1960-1980

Industry	*1980*		*1970*		*1960*	
Agriculture, forestry, fishing, mining	457	(0.2)	1,832	(0.8)	735	(0.4)
Construction	6,664	(3.3)	12,855	(5.5)	10,314	(5.9)
Manufacturing	39,404	(19.6)	60,298	(26.0)	50,886	(28.8)
Transportation, communications, public utilities	19,251	(9.5)	15,261	(6.6)	11,715	(6.6)
Wholesale and retail trade	29,107	(14.5)	35,391	(15.2)	28,247	(16.0)
Finance, insurance, and real estate	12,507	(6.2)	8,646	(3.7)	3,187	(1.8)
Services	74,504	(37.0)	74,874	(32.3)	56,924	(32.3)
Government	19,483	(9.7)	23,035	(9.9)	14,541	(8.2)
Total[a]	201,377	(100.0)	232,192	(100.0)	194,506	(100.0)
Percentage of city total	32.2		30.4		24.7	

SOURCE: U.S. Department of Commerce, Census of Population, *Detailed Characteristics* (selected years).
NOTE: Numbers in parentheses represent intrayear percentages.
a. Data for 1960 are adjusted for 17,958 individuals who did not report the industry in which they were employed.

erous cities. For most older cities like Philadelphia, this meant a disproportionate share of minority clients. With the establishment of public service employment (PSE) programs in the 1970s, local governments extended their employment activities to include direct hiring of the economically disadvantaged and unemployed. For most large city governments, and Philadelphia was no exception, PSE enabled them to hire relatively large numbers of residents, including many minorities.[13] In effect, the local government became the "employer of last resort" during a time of deep recession.

Two other factors reinforced for minorities this embryonic responsibility of local governments. One is the existence of federal equal employment opportunity legislation—discussed above—and the other the emergence of a dominant minority coalition in Philadelphia local politics. With Title VII of the Civil Rights Act, an Equal Employment Opportunity Act, a federal commission (the Office of Contract Compliance) to oversee both, and the strengthening effect all of this had on local human relations commissions, minority group attention was increasingly directed toward government, specifically to local governments, as allies in the quest for employment. The legislation encouraged racial concerns to be channeled through the local government, despite the fact that the legislation was not invested with significant enforcement capabilities as regards the regulation of hiring practices in the private sector.

Though local governments lacked certain powers, they still held out the possibility of employment and could provide important symbolic support for minority employment concerns. This became particularly salient with the emergence in Philadelphia, and other cities, of black political coalitions. In Philadelphia, W. Wilson Goode took office as the first black mayor of Philadelphia in 1984. Blacks had served on the city council prior to this time, but with the ascendancy of Goode into the mayoral seat, they now had a major voting bloc within the dominant political party (Democrat) and occupied the chief executive position (Beauregard, 1988). In the 1987 local election, the Republican mayoral candidate, Frank Rizzo, siphoned off whites from the Democratic party and, as a result, blacks emerged as the majority of registered Democratic voters.

The nature of black politics has been to encourage black expectations as blacks enter positions of authority. One expectation is that black employment needs will be addressed, usually in the form of the hiring and appointment of more blacks to positions on the municipal payroll. Though Mayor Goode has not been able to expand black employment in city jobs, mainly due to fiscal restraints and thus a shrinking work force, he has made significant shifts in the color of his top- and middle-level administrators. Within the minority community, neither the rhetoric nor the potential has gone unnoticed. Minorities in Philadelphia, as in other cities with minority mayors, expect a friendly reception when they approach the city administration.

In sum, the government has the legal mechanisms to confront discrimination in the construction industry and to impose affirmative action guidelines on private sector employers. Additionally, the local government has been cast as a receptive body for minority demands, one likely to respond positively. Last, with shrinking employment opportunities within the city, but for blacks a stable if not expanding governmental sector, the local government also has come to be viewed as capable of providing employment if private sector opportunities are not sufficient.

TARGETING THE CONSTRUCTION INDUSTRY

While there is a certain logic to directing minority employment demands toward the local government, that still does not explain how the construction industry became the target rather than some other industrial sector within the city. To answer this question, we must look more closely at the industrial shifts occurring in Philadelphia during the postwar period, particularly as they affected minorities. In turn, we must view this in light of discrimination, and the attacks upon it, which has been a traditional part of the construction industry.

Philadelphia's industrial structure, as represented by relative employment shares of city residents, has experienced the now-familiar manufacturing to service shift. In 1950, 35.7% of residents were employed in manufacturing. By 1980, that figure was 20.9%. At the same time, service employment went from 18.6% to 32.7%. Resident employment shares also increased in government, finance, insurance, real estate, and wholesale trade. This represents, as numerous researchers have pointed out (Bluestone and Harrison, 1986; Kasarda, 1983), not just a shift in the employment demand generated by different sectors and a change in the "nature" of the economy (from goods production to service delivery), but also a shift in the income distribution as high-wage manufacturing jobs are replaced by lower-wage service jobs. Additionally, the manufacturing-service shift also implies a movement away from full-time unionized, or at least relatively stable, jobs to part-time nonunionized jobs along with a concomitant shift from male to female employment.

For Philadelphia residents, then, the aggregate job structure was shifting away from stable high-paying jobs open to minimally skilled and skilled individuals to lower-paying jobs with unstable prospects requiring almost no additional skills beyond basic reading and writing. As Table 5.1 shows, black workers were experiencing these changes directly. From 1960 to 1980, they have been concentrated increasingly in finance, insurance, and real estate; transportation, communications, and public utilities; and government; while losing shares in manufacturing. Moreover, blacks have been underrepresented in the postwar period in Philadelphia's manufacturing sector, but have held a fairly representative proportion of jobs in services and government.[14] In all sectors, they are increasing their shares. Thus while the manufacturing-service shift is eroding well-paying job opportunities for all residents, blacks are benefiting from the expanding service and government sectors, and even doing well, relative to whites, in manufacturing.

Such data suggest that blacks might do best by joining industrial unions and by organizing in various service industries and in government to assure less discrimination and to create more bargaining power over wages and work conditions. Still, it is also logical for blacks to search for "strong" industrial sectors paying relatively high wages and where discrimination has previously prevented minorities from receiving their share of employment opportunities. This brings us to the construction sector.

Construction jobs within the city have exhibited a secular decline since the early 1950s. Average yearly employment hovered in the low 30,000s in the 1960s but in the middle to low 20,000s in the 1980s.[15] (Table 5.2 exhibits this long-run erosion of employment along with the cyclical sensitivity of the industry.) In addition, the ratio of construction to manufacturing jobs within the city has been increasing. Construction and manufacturing, as will be discussed more fully below, are high-wage sectors for low- or semiskilled workers. For such workers, actual employment opportunities are decreasing in manufacturing, but relative opportunities are increasing in the construction industry.

Overall, resident employment in the construction industry has also declined, from 46,294 in 1950 to 22,921 in 1980. Across the metropolitan area, a similar drop in construction employment has occurred, though not as precipitous. The decline over this time period was from 93,716 to 88,186 (6.3% versus 50.5% for the city).[16] Philadelphia construction workers, in fact, are commuting outside the city for employment, though this is not as true for black construction workers.[17] In effect, the construction industry, particularly within Philadelphia, is not an arena of expanding employment opportunities.

The situation is no different for blacks, and even worse. While the black share of city resident construction employment has fallen slightly (from 31.0% to 29.1%) between 1960 and 1980, the absolute number of black residents employed in construction has declined by 3650, or 35.4%. The decline from 1970 to 1980 is even more dramatic. (See Table 5.1.) These shifts have occurred simultaneously with a 6% black increase in aggregate resident employment over the two decades. In sum, construction jobs within the city are in secular decline, the number of black residents working in construction is falling, and the black share of resident construction employment is virtually stable.

TABLE 5.2
Construction Employment in Philadelphia, 1946-1984

	Total Construction Employment		*As Percentage of Manufacturing Employment*	
Date[a]	*Trough*	*Peak*	*Trough*	*Peak*
1946	22,372		7.0	
1953		39,204		10.9
1962	25,428		8.7	
1967		34,806		12.4
1972	29,220		13.2	
1974		40,005		20.4
1977	21,299		13.7	
1979		26,412		16.9
1984	20,188		17.6	

SOURCE: U.S. Department of Commerce, *County Business Patterns* (selected years).
a. The dates were selected based on an analysis of construction employment cycles from 1946 to 1985. The manufacturing data, however, do not show corresponding cycles but rather an almost straight-line decline from a 1953 peak.

Nonetheless, a number of factors operate against dismissing the construction industry as a target of grass-roots demands for minority employment. Internal to the construction industry are three factors that deserve attention: the amount of investment flowing into construction within Philadelphia, the relative wage rates of construction workers, and the employment opportunities for low-skilled individuals. First, despite postwar population loss, employment decline, and a significant loss of business establishments across all sectors, the flow of investment into construction has not diminished.[18] In constant dollars, both new construction and overall building activity (which includes additions and alterations) have been secularly stable since 1950 with significant peaks occurring after 1966. From 1944 to 1986, the total value of building activity in 1958 dollars averaged $141.7 million per year, with a low of $9.2 million in 1944 and a high of $233.2 million in 1970. In the mid-1980s, in fact, local experts were declaring a building boom in the city as office tower after office tower emerged from the ground in Center City, large public works projects (e.g., a new convention center and a major highway extension) were under way, and residential construction was robust in various sections of the city. During 1986, the value of building construction in current dollars reached $630.8 million, $176.3 million in 1958 dollars. The construction industry seemed to be doing well despite some volatility and despite declining employment.

Second, partly due to widespread unionization, wages in the construction industry are above the metropolitan average. In 1986, the mean annual construction wage was $22,906, while it was $13,542 in services and $20,399 in government. Construction wages greatly exceeded wages in the types of service jobs in which low-skilled minorities were likely to be found: $11,198 in retail trade, $11,527 in hotel and lodgings, and $7,154 in eating and drinking establishments (Naroff, 1987). Of course, construction work is both seasonal and unstable. Still, during slack periods, construction workers can often find alternative employment. The construction industry not only offers higher than average wages but is also accessible to individuals with few skills. Much construction work requires basic, unskilled labor: individuals who can dig trenches, haul brick, strip forms, and paint walls. Certainly, it also requires many skilled individuals, such as cabinetmakers, electricians, steelworkers, and crane operators. Nonetheless, significant segments of construction employment can be filled by unskilled minorities. Thus the "health" of the construction industry, the wage advantages, and the potential for the hiring of unskilled individuals operate as attractions to minority political mobilization.

Joining these internal attractions, and further shifting attention away from declining employment opportunities, is the quarter century of governmental concern with discrimination in the construction industry and the increasing involvement of local governments in private sector construction projects. As noted above, even prior to the passage of the 1964 Civil Rights Act, the construction industry had been selected for scrutiny by the federal government, specifically those construction projects with federal funding. In addition, civil rights activists in the 1960s also targeted the local construction industry for the barriers that construction unions posed to minority employment (Marshall et al., 1978: 26-61). What made the construction industry even more salient in the 1970s was the enlargement of local governmental participation in construction projects.

For centuries, local governments in the United States have been involved in a wide range of public construction projects ranging from roads to market houses to ship channels to school buildings (Benjamin, 1984). Capital improvements have become, in fact, a significant portion of local governmental expenditures.[19] Prior to the postwar period, however, local governmental involvement in private construction projects was ad hoc and relatively rare. With the passage of urban renewal legislation in 1949, local governments became "partners" in private projects on a more institutionalized and systematic basis

(Fainstein et al., 1986). Redevelopment authorities purchased land, demolished buildings, provided infrastructure, and invested in site improvements in order to facilitate the construction of private office buildings, apartment houses, and retail malls. Even with the restructuring of the urban renewal program by the 1974 Housing and Community Development Act, such public-private ventures continued. A few years later, Urban Development Action Grants were instituted and, with the emergence of a wave of redevelopment in the late 1970s and early 1980s, local governmental involvement deepened. By the mid-1980s, not only were many private construction projects in large cities utilizing federal and state funds (e.g., UDAGs, Community Development Block Grants, industrial development bonds), but local governments were also heavily involved in providing such subsidies as property tax abatements, lease arrangements, and employment and training assistance. In some cities—Hartford is a good example (Clavel, 1986: 19-56)—the local government even went so far as to take equity positions in certain property development projects, while cities such as San Francisco and Boston were establishing "linkage" programs requiring contributions by developers to community programs in return for development rights (Keating, 1986). In many cities across the country, a good deal of construction activity could not be easily classified as public or private, but rather included financing and other contributions from both governmental and private sources. Construction is one industry in which the local government has become a significant market participant.

LOCAL GOVERNMENTAL COMPLICITY

While these various factors might explain why the construction industry and the local government became the target for employment demands by minority organizations, still unexplained is why local governments were willing to be placed in such a position. Obvious from the experience of cities with hiring preference ordinances is that the local government allowed itself to become an accomplice in this attack upon the employment prerogatives of private sector firms. Central to the explanation is the above-mentioned transition of local governments from simple contractors of construction services for public capital improvements to financial, developmental, and equity participants in private and public-private ventures (Stone and Sanders, 1987). When combined with local governments' complicity in equal employment legislation and activities and the ascendancy of black elected officials to mayoral and councilmanic positions, the difficulties of claiming no responsibility for discrimination in the construction industry become profound.

The pressure brought upon local governments, however, goes beyond a demand for the end to discrimination. It also concerns an affirmative action response to the low levels of minority employment, and thus includes a demand for an expansion of job opportunities. The employment side of these controversies, though inextricably linked to discrimination, intersected with another change in local governmental activities to provide an additional reason for local government participation.

By the 1980s, "jobs" had become a powerful symbol of local governmental performance. Elected officials across the country touted their past accomplishments in terms of employment growth, and polished their futures by claiming to make jobs their number one priority (see Bergman, 1986: 1-14). The postwar period had witnessed an unquestioned expansion in governmental interest in local economic development and employment and training activities. The federal government had taken the initial lead. By the late 1970s and early 1980s, the Carter and then Reagan administrations were withdrawing the federal government from subnational economic development, while recessions in the mid-1970s and early 1980s caused local governments to give more and more attention to economic development activities. Local governments expanded their economic development agencies, and elected officials and administrators cast their efforts in terms of jobs. The call for hiring preference ordinances played well within that perspective. While it did not require the local government actually to create jobs, it did require that local residents be given priority for employment opportunities existing within the locality. The demands, moreover, linked equal employment opportunity, the dominant economic development interest of local government, and the participant status of local government in the local construction industry. A powerful argument had been created.

Nonetheless, certain political and legal consequences attendant to the passage of hiring preference ordinances tend to discourage local governmental involvement. Local construction unions and business interests frequently oppose these ordinances and, because of their numbers and importance to the city, wield a good deal of political influence. In addition, the ordinances are susceptible to legal challenge (Beauregard, 1987; Clark, 1985: 85-110). The Boston ordinance, as noted above, successfully withstood a challenge based upon the Commerce Clause. Still, the city had to expend funds on a very complex and time-consuming legal defense. In Camden, on the other hand, the plaintiffs (the United Building Construction and Trade Council) took a different legal tack. Their case was built upon the Privileges and Immunities Clause, which prohibits subnational governments from making any law that denies to a nonresident the rights and privileges that person has under other jurisdictions, mainly other states. The U.S. Supreme Court agreed with the legal position of the plaintiff and remanded the case back to the New Jersey Supreme Court for fact-finding. For other cities, the situation is clear: These ordinances are susceptible to a challenge based upon the privileges and immunities clause. Even though the challenge focuses on the resident goal, and not the minority or female goal, the end result is still an illegal ordinance.

When rendering its decision in *UBCTC v. Camden*, the U.S. Supreme Court also raised the specter of a balkanized United States with each locality attempting to control employment opportunities within its boundaries. Mayor Goode in Philadelphia used this idea to rationalize his opposition. As a city with declining employment opportunities and with investment in construction robust but still not unchallenged by suburban construction investment, Philadelphia ostensibly could not afford to have surrounding municipalities legislate similar ordinances. The soundness of this position is reinforced by commuting data, which show Philadelphia construction workers increasingly finding employment outside the city. Other, older industrial cities display similar characteristics. Thus there are significant reasons for local governments to avoid complicity. Nonetheless, in numerous instances, local governments have entertained the possibility of, if not actually passed, hiring preference ordinances.

THE SPATIAL DIMENSION

Finally, the spatial dimension of many of these factors should not remain implicit. The debate over hiring preference ordinances is made possible because of the spatial nature of economic restructuring, of the social restructuring of cities, and of property development.

While few areas of the country have been spared the manufacturing-service shift, its unfolding has been uneven. Philadelphia, for example, has had both a relative loss and an absolute loss of jobs in manufacturing, while Houston has had a decline in manufacturing's relative share with an absolute increase (Beauregard, forthcoming). Moreover, when Philadelphia was losing high-wage manufacturing employment and substituting low-wage service employment, its suburbs were not doing as badly in manufacturing, doing better as regards services, and, in the 1980s, doing well in high-wage research and development and office industries. Simultaneously, the high-wage jobs being generated in Philadelphia's office industries were more likely to be taken by commuters than residents. These types of spatially disaggregated changes make the balkanization argument less powerful. The national economy is already characterized by uneven spatial development; growth and decline are concentrated in nodes, while the growth centers of "distressed" central cities leak jobs to surrounding municipalities. Under these conditions, the need for resident hiring preference ordinances in older, central cities becomes more salient.

The social landscape is equally differentiated. The metropolitan area's minority population is concentrated in Philadelphia and in two depressed cities: Camden, New Jersey, and Chester, Pennsylvania. The school system in Philadelphia is one of the weakest in the metropolitan area, educational levels are lower, and unemployment rates are higher. The low-skilled, poorly educated labor force is disproportionately contained within the central city, creating one side of the often-mentioned "mismatch hypothesis" (Kasarda, 1983). The population in most need of governmental assistance and advocacy for any expansion of employment opportunities is the minority population of Philadelphia. While one could argue that the solution is simple—improved metropolitan mass transit—that solution is unlikely to be successful given the social and economic costs of commuting in comparison with the wages paid in the types of jobs—mostly service—minority workers would likely find in the suburbs (see Bederman and Adams, 1974). Racial segregation thus plays a role in the emergence of hiring preference ordinances.

Last, the rediscovery of the central city as an "administrative-service center" for the region has helped to concentrate construction investment in Philadelphia.[20] A spatial centrality is being reasserted even while decentralization continues on a number of dimensions. This is important because property development is spatially fixed. The places of value that attract new investment cannot be relocated as one would textile machinery. Property investment is "frozen" not "footloose." The activities that occupy the property can be moved to sites outside of the city, but only *after* the construction phase has ceased. To this extent, the construction industry is somewhat captive within the locality, giving the local government additional leverage.[21] The demand for hiring preference ordinances, then, cannot be understood outside of its spatial context.

CONCLUSIONS

Minorities in Philadelphia in the last few decades have been faced with a shrinking number of employment opportunities in the local labor market. Despite a corresponding decline in construction employment, including a sharp drop-off in minority participation, blacks turned to that industrial sector for jobs. Jobs in construction have expanded relative to manufacturing within the city, investment in construction activity has remained stable but with numerous boom periods, and construction wages are high when compared to those in other local industries. In terms of skill requirements, construction jobs are accessible to blacks with little formal education. Industrial restructuring and the construction sector's own characteristics suggest that this sector would be an appropriate source of minority jobs, and the discrimination that has pervaded construction as well as the existence of equal employment laws provide a rational for involving the government.

The government became a target for minority demands because of its involvement in antidiscrimination activities, because a significant black electoral coalition emerged, and because local governments became significant employers, entered into private property development, and claimed responsibility for job generation within their boundaries. Despite constitutional limitations, these factors together make the local government vulnerable to minority demands for hiring preference ordinances. The various forces, in addition, take much of their strength from the spatial transformations of postwar United States that have concentrated urban decline and minorities in central cities. At the same time, the spatial fixity of construction activity has further strengthened minority demands.

Economic restructuring established the context for political mobilization around hiring preference ordinances by creating high levels of unemployment and underemployment in the minority community, elevating the construction industry to high visibility, and providing one inducement for local governmental involvement. Political restructuring made the local government a legitimate source of redress for minority grievances and a market participant in property development, and thereby established the specific lines of action for the unfolding of these political events. Targeting the construction industry, in fact, only makes sense in economic *and* political terms. The restructuring of the local economy did not make construction the only option, but the involvement of local government in construction when paired with federal equal employment laws highlighted it as a political target. Moreover, the local government is vulnerable. It cannot easily dismiss minority demands for intervention into the employment relation, not when it is heavily implicated in the generation of jobs and the management of labor. Thus, while we witness here the strong influence of economic restructuring, the political response is not a direct outcome. Rather, it is a consequence mediated, in clear fashion, by political and spatial restructuring.[22]

OVERCOMING THEORETICAL DIVISIONS

From the 1930s to the late 1970s in the United States, the federal government increased its involvement in the reproduction of the labor force, and state and local initiatives that had arisen during the Progressive Era were diminished in importance. Postwar prosperity was particularly important in creating a strong national government with broad interests and was instrumental in establishing a social contract between labor and capital that effectively dispelled any need for direct and extensive governmental mediation of workplace struggles. Thus given a protective welfare state on the one side and a well-functioning economy on the other, and labor reaping the benefits of both, it is not surprising that Marxist political economists turned their attention to the state and pushed labor issues to the background. Even as the state was being recognized for its contributions to accumulation (O'Connor, 1973), the recognition of industrial capitalism's spatial separation of workplace from home and advanced capitalism's severance of production from reproduction redirected theoretical investigations toward issues of collective consumption, state behavior, and related political movements. Capitalist prerogatives, and thus state limits, were seemingly clear: the state "has no authority to order production or to control it" (Offe, 1975: 126).

At the same time, the ability of capital to be spatially mobile was viewed as a force that shifted political involvement in the accumulation function to the national level. Because state and local governments could only rearrange accumulation, and because any intervention by them into capital investment or the labor process that hindered accumulation would simply result in capital's flight, the local government's participation in the sphere of production was practically and theoretically untenable.

As a result, Marxist political economists, particularly those who concern themselves with local politics and other "urban" issues, are relatively quiet as regards labor issues. Most Marxist urban theory has not only taken its flavor from the rise of the national state but has also reacted strongly to the spatial separation of workplace and home. The latter is particularly important in thinking through political behavior at the local level. Castells (1977, 1983), for example, uses the rise of the state and the spatial separation of residence from workplace to direct our attention to the increasingly collective nature of reproduction and the shift of political movements to state channels. The local government is conceived as almost wholly preoccupied with collective consumption. Katznelson (1981) takes the spatial separation and reformulates it as a political division in which labor issues are fought out at the workplace, often through struggles between organized workers and capitalists, while issues of reproduction and consumption are fought out through local governments and political parties and between citizens organized on the basis of place and ethnicity. Last, Fainstein and Fainstein (1985: 190) list three principal forms of conflict between the state and urban social movements: communal demands for territorial self-management, calls for more or better collective consumption goods, and issues of communal defense. Only in passing (p. 193) do they mention affirmative action in employment in their discussion.

Marxist urban theorists thus arrive at an interpretation of advanced capitalism that distances the state, for very good theoretical reasons nonetheless, from workplace concerns. Lacking is both a recognition that economic and political restructuring can bridge the "city trenches" with social action, and an attempt to build a theoretical framework that would connect workplace with home, production with reproduction, and state behavior with the social relations of production. The holism that is supposedly a hallmark of Marxist thinking is thereby being undermined by capitalism as it creates divisions within society that channel lines of investigation, and by urban theorists who frequently slip across that fine line between focus and myopia. Marxist labor theorists (Braverman, 1974; Gordon et al., 1982) are equally constrained by these new disciplinary boundaries and have neglected the realm of reproduction and the state's involvement in linking production and reproduction concerns, and workplace with home.

The issue discussed here—hiring preference ordinances—demonstrates a clear link between local politics, the employment relation, and the local government. The local government finds itself intervening in the labor process. Collective consumption is not the issue, but rather the relation between reproduction and production. A workplace concern is being debated and resolved outside the workplace. This seems to be a challenge to the general Marxist understanding of state behavior and urban political movements under advanced capitalism. Hiring preference ordinances and equal employment laws, then, should cause us to rethink the present divisions within society and within Marxist urban theory. Economic and political restructuring are unearthing novel political issues, redefining the relation between capital and the state, and affecting the nature of local governments. The end result is a new series of contradictions and tensions emanating from the capitalist social formation and confronting the complacency of theorists.

The overall thrust of recent history seems to be toward greater involvement of the state, and particularly the local state, in elements of the labor process and in certain aspects of investment and production. In part, what we might be seeing is an overreaction of U.S. capital and the state to a loss of economic and political hegemony in international markets and diplomatic circles, to the relative diminution of manufacturing, and to the rise of advanced service industries whose profits are tightly linked to state-controlled money supplies, interest rates, and debt policies. The turn of the federal government away from welfare state activities and subnational economic assistance, thus forcing local governments into more aggressive economic roles, too, might well pass. Even if these trends prove to be temporary, the relations between capital and the local state will not likely return to their original positions.

At present, property capital seems to have traded less economic and financial risk for more political risk. By engaging in public-private partnerships and by allowing local governments to become more directly involved in different phases of the property development process, property capital has found another "investor"—one that often provides equity without taking an equity position—but has also made that new participant a conduit for the political demands of various oppressed groups within the community. By subsidizing construction, for example, it becomes difficult for local governments to deny the demands of residents, minorities, and women for equal employment opportunities on construction projects. Property capital thus finds itself the target of intense public scrutiny.

Industrial capital, on the other hand, here represented by the construction industry, seems to be trading off control over the employment relation for better protection from direct political attack. Whereas earlier demands for minority hiring on construction projects led to demonstrations and even violence on construction sites, the arena of conflict has been shifted to councilmanic chambers. Minority groups now utilize equal employment laws to legitimize demands directed at politically sensitive project participants, the local governments. The state has expanded its support of investment and accumulation, but that support has fostered a rethinking of the employment relation, thus undermining capital's control over production.

The local government has not, however, acted alone and on its own initiative. The withdrawal of the federal government from subnational economic development policies, the emergence of state-subsidized urban redevelopment projects, and the increasing hypermobility of capital have engendered intergovernmental competition for investment. This has virtually forced the local government to be more aggressive as regards investment and capital accumulation within its midst. Local governments are now "partners" in economic development and property investment.

Attempts to capture investment within municipal and state boundaries balkanize the national economy and pit local governments against both the Constitution's concern for national economic integration and the federal government's concern for national political integration. Hiring preference ordinances cause problems on each of these dimensions.

At the same time, and as a result of their new role, local governments establish a different relationship with accumulation and legitimization. More direct involvement in accumulation means that legitimization of the state as the preferred channel for political activity can easily lead to demands that the state's accumulation activities be used to satisfy a broad range of concerns. Profits earned by local government as a market participant would then become resources for satisfying political demands unrelated to the original investments. Thus, because of its involvement with property development, local government ultimately intervenes in the employment relation within the construction industry and reproduces the legitimization instilled by civil rights legislation. In addition, that intervention threatens capitalist control over the workplace. As a consequence, the local government loses some of its relative autonomy vis-à-vis capitalism. The state is not gaining control over capital by becoming a market participant. Rather, it becomes even more susceptible to the logic of capital as public funds are invested in profit-making ventures.

City residents, finally, also find themselves faced with a complex set of gains and losses from contemporary economic and political restructuring. While citizens have gained an ally in their quest to control the employment relation, their ally is suspect and the newfound influence possibly transitory. The state might mediate between labor and capital, but it comes to the negotiating table with a distinct bias. The state under capitalism can never sever totally its need for and its structural subservience to capital. Moreover, even though forays into the legal realm enable labor to exploit the indeterminacy of the Constitution and of legal precedence (Clark, 1985: 35-59), the isolated successes of labor are not to be confused with the structural inclination of the law toward private property and the prerogatives of capital. Within the employment relation, this means that the limits of political action ultimately stop at the issue of who owns the jobs. Labor's new ally in the struggle over the labor process also bears another cost: Energy expended on developing and implementing political strategy to pressure the local government means less energy directed toward the workplace. Labor thereby loses influence over capital as it channels its political demands to the state.

In sum, one cannot allocate reproduction issues to the state and production issues to the workplace. One cannot assume that the spatial separation of work from home is also a political division. One cannot discount the importance of the local state in employment issues, and one cannot reflect economic restructuring onto the social backdrop and expect to read from it the nature of local political activity. The restructuring of the economy certainly includes a restructuring of the relation of capital to the state, but the state is also undergoing its own restructuring. Changes in civil society, moreover, bear down on all of these relationships. In the interstices, opportunities for collective political action are being demolished, while others are being etched on the landscape. Minority labor has discovered a channel for addressing the employment relation, one that not only provides the potential for well-paying jobs but one that also encapsulates the unique transformations currently under way in the U.S. political economy. The challenge is to theorize the uniqueness of political action within the constantly evolving structure of capitalist society.

NOTES

1. Plant closures are another issue that leads to community groups confronting local governments in quest of state intervention into the labor process. See Bluestone et al. (1981), Lauria (1986), and Metzgar (1980). There have also been instances where local governments have attempted to intervene in the investment decisions of corporations, usually to prevent a relocation outside of the city or environmental degradation. See various articles in Kelly and Webb (1979).

2. Of course, the National Labor Relations Act explicitly rejects the assumption of the freedom-of-contract doctrine that individual laborers enter contract negotiations with capital as equals (Clark, 1986).

3. On the other hand, "the areas where blacks had made the most significant gains were the very industries—shipbuilding, aircraft, munitions, and explosives—with the least potential for reconversion to peacetime production. . . . Almost one-half of the blacks in war industries had been concentrated in aircraft, shipbuilding, iron and steel, and government jobs related to war services" (Fusfeld and Bates, 1984: 73).

4. For a good overview of the sequence of equal employment legislation and a lament as to its shift from individual rights and nondiscrimination at hiring to group rights and an emphasis on outcomes, see Glazer (1975: 33-76). For Glazer (p. 75), equal employment laws have produced a nation "increasingly divided formally into racial and ethnic categories and differential rights."

5. "If anything, few other industries afford minorities relatively as many high-paying opportunities as the building trades" (Marshall et al., 1978: 11).

6. The Philadelphia Plan, instituted by the OFCC, was one of the first to require goals and timetables. It faced a legal challenge by the Contractors Association of Eastern Pennsylvania and was upheld in 1970 by a federal district court, which maintained that affirmative action was a valid exercise of presidential powers. Ironically, the court claimed that while goals were legal, an attempt to force the employer to achieve those goals, thus turning them into quotas, might be illegal (Marshall et al., 1978: 10).

7. The Massachusetts Council of Construction Employers also challenged the ordinance on two other grounds: (1) the Privilege and Immunities Clause, which prohibits discrimination by one state against the citizens of another state, and (2) the National Labor Relations Act, which establishes contractual obligations between unions and contractors and whose federal statutes override state actions (Clark, 1985: 98-103).

8. The case was heard by the U.S. Supreme Court whose majority opinion found the ordinance in violation of the Privileges and Immunities Clause—not the Commerce Clause—and remanded the case to the New Jersey Supreme Court for fact-finding. At that time, negotiations produced a nonlegal resolution.

9. The description of the evolution of the bill is based upon local newspaper articles, interviews with people knowledgeable about the controversy, and attendance at various communitywide meetings on Bill 649. See Beauregard (1987, 1988).

10. Data are from the U.S. Department of Commerce's *County Business Patterns* for the selected years. The City of Philadelphia is coterminus with Philadelphia County.

11. Unless otherwise noted, data in the text are from the decennial census of the U.S. Department of Commerce.

12. These data are from the Commonwealth of Pennsylvania, Department of Commerce, *Pennsylvania Statistical Abstract* for various years. In 1983, the local government had 33,419 employees, and 32,421 full-time equivalents.

13. To a degree, a major component of the decline in local government employment from the 1970s to the 1980s was the cessation of federal PSE programs in the early 1980s.

14. In 1960, blacks constituted 26.7% of the city's population and 19.4% of city's employed workers in manufacturing, 36.1% of employed workers in services, and 30.4% of employed workers in government. In 1980, the corresponding percentages were 37.8 of the population, 30.2 in manufacturing, 36.5 in services, and 37.6 in government.

15. Data were taken from U.S. Department of Commerce's *County Business Patterns* (selected years).

16. Construction employment data pose a problem when using the decennial census; construction employment is cyclically sensitive and census dates are often at incompatible points along the business cycles. For a sense of this, compare the census dates with the cycles indicated in Table 5.2.

17. In 1970, 19.7% of the city's construction workers were employed outside Philadelphia, whereas, by 1980, this had increased to 21.3%. For black Philadelphia construction workers, the corresponding percentages were 11.1 and 11.1. Data are from U.S. Department of Commerce's *Journey to Work* for 1970 and 1980, but must be carefully interpreted. To wit, in any one year, construction workers will probably be employed on multiple projects located in more than one municipality.

18. In addition to a 23.2% decline in nongovernmental employment between 1953 and 1985, Philadelphia lost 29.1% of its business establishments over that time period and 18.5% of its population between 1950 and 1980. The employment and establishment data are from *County Business Patterns* (U.S. Department of Commerce). The data on the dollar value of building activity were collected from the City of Philadelphia's Department of Licenses and Inspections as part of a research project funded by the National Science Foundation (NSF-SES 86-07890).

19. For the city of Philadelphia, capital outlays as a percentage of total expenditures increased from 11.5 in 1973 to 14.9 in 1982 with a high of 17.2 in 1979. Data are from *Pennsylvania Statistical Abstract* for various years.

20. Between 1970 and 1984, Center City office space increased by 65% as 36 new buildings (15.6 million square feet) were added. See Philadelphia City Planning Commission (1985).

21. Obviously, different types of capital have different degrees of mobility, and construction investment when it functions as a derived demand is dependent on more mobile capital.

22. To state, as Mingione (1981: 34) does, that "it is impossible to establish a fixed typology of relations between class struggles and the accumulation process, on the one side, and the state and political actions and organizations on the other" seemingly denies the possibility of being able to identify the key factors that define any political event.

REFERENCES

BEAUREGARD, R. A. (1987) "Resident hiring preference ordinances: a comparative analysis." Economic Development Quarterly 1 (2): 124-135.

BEAUREGARD, R. A. (1988) "Tenacious inequalities: race and politics in Philadelphia." Department of Urban Planning and Policy Development, Rutgers University. (unpublished)

BEAUREGARD, R. A. (forthcoming) "Urban restructuring in comparative perspective," in R. A. Beauregard (ed.) Atop the Urban Hierarchy. Totowa, NJ: Rowman & Littlefield.

BEDERMAN, S. H. and J. S. ADAMS (1974) "Job accessibility and unemployment." Annals of the Association of American Geographers 64 (3): 378-386.

BENJAMIN, R.L.C. (1984) "From waterways to waterfronts: public investment for cities, 1815-1980." In R. D. Bingham and J. P. Blair (eds.) Urban Economic Development. Beverly Hills, CA: Sage.

BERGMAN, E. [ed.] (1986) Local Economies in Transition. Durham, NC: Duke University Press.

BLUESTONE, B. and B. HARRISON (1986) "The great American job machine: the proliferation of low wage employment in the U.S. economy." Prepared for the Joint Economic Committee, U.S. Congress, Washington, DC.

BLUESTONE, B., B. HARRISON, and L. BAKER (1981) Corporate Flight. Washington, DC: Progressive Alliance.

BRAVERMAN, H. (1974) Labor and Monopoly Capital. New York: Monthly Review Press.

BURSTEIN, P. (1985) Discrimination, Jobs, and Politics. Chicago: University of Chicago Press.

CASTELLS, M. (1977) The Urban Question. Cambridge: MIT Press.

CASTELLS, M. (1983) The City and the Grassroots. Berkeley: University of California Press.

CHAFE, W. H. (1986) The Unfinished Journey. New York: Oxford University Press.

Citizens' Commission on Civil Rights (1984) Affirmative Action to Open the Doors of Job Opportunity. Washington, DC: Center for National Policy Review.

CLAQUE, E. and L. KRAMER (1976) Manpower Policies and Programs. Kalamazoo, MI: W. E. Upjohn Institute.

CLARK, G. L. (1981a) "Law, the state and the spatial integration of the United States." Environment and Planning A 13: 1197-1232.

CLARK, G. L. (1981b) "The employment relation and spatial division of labor: a hypothesis." Annals of the Association of American Geographers 71 (3): 412-424.

CLARK, G. L. (1985) Judges and the Cities. Chicago: University of Chicago Press.

CLARK, G. L. (1986) "Restructuring the U.S. economy: the NLRB, the saturn project, and economic justice." Economic Geography 62 (4): 289-306.

CLARK, G. L. and M. GERTLER (1983) "Local labor markets: theories and policies in the U.S. during the 1970s." Professional Geographer 35 (3): 274-285.

CLARK, G. L., M. GERTLER, and J. WHITEMAN (1986) Regional Dynamics. Boston: Allen & Unwin.

CLAVEL, P. (1986) The Progressive City. New Brunswick, NJ: Rutgers University Press.

COLLINS, H. (1986) "The Philadelphia plan, 17 years later." Philadelphia Inquirer (January 19): 1-A, 18-A, 19-A.

Commonwealth of Pennsylvania (selected years) Pennsylvania Statistical Abstract. Harrisburg, PA: Department of Commerce.

DUBINSKY, I. (1973) Reform in Trade Union Discrimination in the Construction Industry. New York: Praeger.

FAINSTEIN, S. S. and N. I. FAINSTEIN (1985) "Economic restructuring and the rise of urban social movements." Urban Affairs Quarterly 21 (2): 187-206.

FAINSTEIN, S. S. et al. (1986) Restructuring the City. New York: Longman.

FUSFELD, D. R. and T. BATES (1984) The Political Economy of the Urban Ghetto. Carbondale: Southern Illinois University Press.

GLAZER, N. (1975) Affirmative Discrimination. New York: Basic Books.

GORDON, D. M., R. EDWARDS, and M. REICH (1982) Segmented Work, Divided Workers. London: Cambridge University Press.

KASARDA, J. D. (1983) "Entry-level jobs, mobility, and urban minority unemployment." Urban Affairs Quarterly 19 (1): 21-40.

KATZNELSON, I. (1981) City Trenches. Chicago: University of Chicago Press.

KEATING, D. (1986) "Linking downtown development to broader community goals." Journal of the American Planning Association 52 (2): 133-141.

KELLY, E. and L. WEBB [eds.] (1979) Plant Closings. Washington, DC: Conference on Alternative State and Local Policies.
LAURIA, M. (1986) "Towards a specification of the local state." Antipode 18 (1): 39-63.
LEVITAN, S. A., W. B. JOHNSTON, and R. TAGGART (1975) Still A Dream. Cambridge, MA: Harvard University Press.
MARSHALL, R. et al. (1978) Employment Discrimination. New York: Praeger.
METZGAR, J. (1980) "Plant shutdowns and worker response." Socialist Review 10 (5): 9-49.
MINGIONE, E. (1981) Social Conflict and the City. Oxford: Basil Blackwell.
NAROFF, J. L. (1987) Regional Economic Trends: Summer. Philadelphia: Fidelcor.
O'CONNOR, J. (1973) The Fiscal Crisis of the State. New York: St. Martin.
OFFE, C. (1975) "The theory of the capitalist state and the problem of policy formation." In L. N. Lindberg et al. (eds.) Stress and Contradiction in Modern Capitalism. Lexington, MA: Lexington.
Philadelphia City Planning Commission (1985) "The Economy of Center City." Philadelphia: Author. (report)
STEINBERG, R. (1982) Wages and Hours. New Brunswick, NJ: Rutgers University Press.
STONE, C. N. and H. T. SANDERS [eds.] (1987) The Politics of Urban Development. Lawrence: University Press of Kansas.
U.S. Department of Commerce (selected years) County Business Patterns. Washington, DC: Government Printing Office.
U.S. Department of Commerce, Bureau of the Census (selected years) Census of Population: Detailed Characteristics. Washington, DC: Government Printing Office.
U.S. Department of Commerce, Bureau of the Census (1970) Journey to Work. Washington, DC: Government Printing Office.
U.S. Department of Commerce, Bureau of the Census (1980) Journey to Work. Washington, DC: Government Printing Office.

6

Spatial Change and Social Justice: Alternative Economic Development in Chicago

ROBERT P. GILOTH
ROBERT MIER

FOR DECADES preceding the middle 1960s, the liberal policy agenda in the United States had captured the imaginations of those who cherished smallness and autonomy—the family, the neighborhood, the church, and the farm. The emergence of huge, often inert, liberal institutions such as the federal bureaucracy and labor unions, and the struggles within the New Left, made tenuous or obfuscated the connection between the liberal agenda and common people's concerns (Lerner, 1988). The national agenda had overwhelmed local agendas, setting the stage for the conservative renaissance.

In 1987 and 1988, a plethora of candidates sought the presidency; yet that race stood in stark contrast to battered urban landscapes of closed factories, abandoned housing, and substantial unemployment. Beyond the candidacy of Reverend Jesse Jackson, there was little discussion about the need to reconnect institutions capable of preserving or advancing civil and human rights with the immediate concerns of people chafing under the often-repressive tendencies of large institutions—a challenge that frequently surfaces on the urban terrain (Jackson, 1987). Rather, the candidates counseled the limitations of local action and

AUTHORS' NOTE: *We thank Wim Wiewel, Diana Robinson, and Bob Beauregard for their thorough and thoughtful criticism of earlier drafts. Like many Chicagoans, we were stunned and saddened by the untimely death of Mayor Harold Washington. His moral principles, historical perspective, and political courage provided inspirational leadership, some small portion of which we hope shows through in this chapter.*

promised a national strategy that confronts the global political economy. To put it another way, the popular viewpoint about urban social and spatial changes sees recent upheaval in inner-city industrial and working-class zones as determined by national and international economic restructuring beyond the influence of local planners and communities (Brandy, 1987; Fainstein and Fainstein, 1987).

We believe the presidential debates, with their focus on the global economy, have overlooked recent local experiences that contain lessons for a new national strategy. We suggest that distinctive local responses to events of economic restructuring such as plant closings show that there are continuing opportunities at the local level to articulate new development ideas and practices, to cultivate new coalitions, and to explore new relations between government institutions and social movements.[1] Further, we suggest that these local experiments, while adding to the growing list of challenging development efforts around the United States, are also nurturing principles of social justice, fairness, openness, and participation as seeds for a national strategy (Bruyn and Meehan, 1987; Clavel, 1986). Yet we argue, in contrast to those who overemphasize the rhetoric of "equity planning," that such community experimentation faces enormous obstacles in the form of growth coalitions, community fragmentation, unresponsive bureaucracies, and intractable local problems (Squires et al., 1987).[2]

We explore the potential of local efforts by examining Chicago's experience between 1983 and 1987—largely the administration of Harold Washington. Although we emphasize distinctive features of the Chicago situation, we are not arguing that Chicago is unique. First, we outline several preconditions for experimentation as it emerged in Chicago. Second, we present three examples that illustrate the strategies in action: the Playskool plant closing, industrial displacement, and the implementation of local industry task forces. In each example, we attempt to connect the Chicago response to similar efforts in other cities. Third, we reflect on this experience for lessons about strategy development, coalition building, community action, replication of responses, and the connection of local actions to national change.

A CLIMATE FOR EXPERIMENTATION

During the period 1983 to 1987, one found an unlikely set of Chicago groups promoting economic development:

environmental activists helping small manufacturers with toxic wastes;
industrial councils of small manufacturers collaborating on worker buy-out studies;
public housing tenant councils working with local businesspersons in starting small businesses;
a social worker leading a task force to set up a capital fund for neighborhood manufacturers; and
union leaders supporting legislation to prevent displacement of industries by downtown development.

These are but a few examples of the creative ferment in Chicago around economic development—much of it focused on how to retain and expand the city's manufacturing base (Greenhouse, 1986; "Help for a Day," 1987). That ferment has several distinctive elements: the diversity of innovations being explored simultaneously; the coalition approach to problem solving; the ongoing difficulty of stretching these coalitions across racial boundaries; and the turn to production-oriented issues such as business investment by neighborhood activists.

Several elements came together in Chicago in the 1980s that created a climate for experimentation in progressive economic development. The first was the maturation of an extensive network of neighborhood organizations interested in economic development, including several specialized organizations within this network dedicated to innovation, technical assistance, and advocacy. The second was the evolution of an alternative set of development ideas and generative metaphors that enabled the framing of problems in new ways and the invention of distinctive solutions. Finally, these ideas and networks expanded dramatically beginning in 1983 with Harold Washington as mayor—networks began to overlap with government institutions and alternative ideas gained new legitimacy as part of Washington's development program.

These networks and ideas generated quite distinctive responses to events commonly associated with economic restructuring—plant closings, industrial land-use conflict, and general industry distress. The events expressed both the negatives of restructuring, but also opportunities for new actors to define new solutions. Three related economic processes also explain the impact of restructuring at the local level: industrial concentration/fragmentation; spatial concentration/dispersal; and a bifurcation of the labor market. Before turning to our case studies, however, these three processes and the preconditions for progressive economic development—networks, ideas, and the administration of a progressive mayor—will be discussed.

COMMUNITY NETWORKS

A rich, differentiated network of organizations and activists concerned with alternative approaches to economic development emerged in Chicago by the early 1980s. There were several key components to this network as of 1980: organizer training centers; networks of neighborhood groups; technical assistance centers; and a growing number of community experiments, ranging from business incubators to fighting industrial displacement.[3] These organizations and experiments were often supported by an enlightened group of local foundations such as the Wieboldt, Joyce, and, later, MacArthur, and the Woods Charitable and Chicago Community trusts.

The policies of President Reagan and the recession of 1981-1983 gave a new impetus to neighborhood groups to develop their thinking about economic development. Facing cutbacks in funding, they began to move away from a sole focus on consumption strategies in the arenas of housing and social welfare (Giloth, 1985; McKnight and Kretzmann, 1984). For example, Reagan's enterprise zone proposal and its attendant rhetoric emphasizing laissez-faire capitalism galvanized a community opposition already mobilizing around human rights. The result in Chicago, in response to the introduction of a state enterprise zone bill, was the creation of another important network of grass-roots organizations, the Chicago Workshop on Economic Development (CWED). Finally, decline in basic industry, epitomized by the steel industry, began to connect labor movement people with community people, thus bringing about the formation of the Midwest Center for Labor Research (MCLR).

This constellation of forces came to a head in late 1982 with the mayoral campaign of Harold Washington. Washington was a candidate around whom neighborhood, labor, and civil rights activists would coalesce. His legislative career was marked by extraordinary sensitivity to the needs and aspirations of those constituencies, and culminated in the leadership role he played as a U.S. congressman in passage of the historic 1982 Voting Rights Act. His mayoral campaign directly appealed to a neighborhood constituency, utilizing their leaders in key campaign roles, and adopting their ideas and their rhetoric. His victory became a mandate for him to bring community leaders into local government and to implement the neighborhood agenda.

Mayor Washington began by working with the networks of development groups to chart a more responsive and equitable approach to development. His *1984 Development Plan* committed the city to a development agenda that was fair in recognizing all Chicago neighborhoods, jobs oriented, and balanced between downtown and neighborhoods, and manufacturing and service industries (Bennett et al., 1987; Judd and Ready, 1986; Krumholz et al., 1985). It cast city government in a role of responding to local initiatives, such as community-created business incubators and worker-owned businesses, by involving citizens in the business of government.

ALTERNATIVE DEVELOPMENT IDEAS

The antinomy of neighborhood versus downtown—a long-standing, urban grass-roots, metaphor—was transformed in Chicago and elsewhere in the 1980s to portray a new set of development choices: manufacturing versus the service economy; blue-collar jobs versus low-wage McJobs; job generation versus real estate development; industrial expansion versus downtown growth; credit-starved neighborhoods versus the growth of the finance industry; targeted local hiring versus regional business climate; and minority/female businesses versus efficiency. These new contrasts resonated with the historical meanings—particularly apt for Chicago—of the downtown/neighborhood metaphor: city hall versus the neighborhoods; the rich and powerful versus the poor and excluded; corporations versus small business; and business as usual versus good government (Barret, 1983).

In the end, this complex of ideas in the black and Hispanic communities and the neighborhood network yielded several new development strategies in Chicago (Mier et al., 1986; Stanback and Mier, 1986-1987). The first opened government and its resources to previously excluded constituencies such as neighborhood organizations, civic groups, and labor unions, while increasing accountability for those firms or institutions who received public incentives for development. The second was the exploration of ways to link redistributively downtown and the neighborhoods, including requirements for minority and female business purchasing and targeted hiring. Finally, an attempt was made to balance development across economic sectors, size of business, and areas of the city. As will be clearer in the case studies, these strategies overlap in practice.

Balanced growth as a strategy to ameliorate urban inequities was rooted in the urban landscape in the late 1970s and early 1980s:

downtown boom, industrial collapse, and neglect by the Reagan administration. At the same time that the U.S. Steel South Works and Wisconsin Steel plants on Chicago's southeast side were closing, laying off 13,000 workers, and devastating nearby communities, a new futures exchange was being completed in Chicago's "Loop" with the help of a $19 million federal Urban Development Action Grant (UDAG). This development spurred South Loop residential development with an upscale loft district and a new-town-in-town.

The pervasive experience of glitter and collapse brought together grass-roots constituencies with a new set of ideas. A long-gestating realignment of neighborhood activism toward economic development and redistributive/accountability strategies was unwittingly spurred by the reduced funding, privatization, and entrepreneurial ideology of Reaganomics. This led neighborhood and civic activists to reformulate notions about development, governance, and their role, often relying upon the downtown/neighborhood metaphor as an analytical tool to design strategies and as an organizing theme (Mier and Gelzer, 1982). Three ideas emerged: Chicago's neighborhoods, small businesses, and manufacturing were being neglected, downtown development and the service industry were overemphasized, and downtown development impacts, particularly the questions of who pays and who benefits, were not being addressed. These ideas joined a strong "reform as fairness" theme that came out of Chicago's black community, particularly during the Washington campaign. That theme was not simply a reworking by a new ethnic group of "we want ours," but basic principles about social justice and fairness (Grimshaw, 1984).

THE ECONOMIC ENVIRONMENT

With alternative development ideas and a network of groups, including city hall, the climate existed for mounting novel responses to economic restructuring. Several economic, sectoral, and spatial processes were at work in this time period that generated opportunities and reinforced the salience of a balanced growth framework: industrial concentration/fragmentation, spatial concentration/dispersal, and bifurcation of the labor market.

Industry concentration/fragmentation. Some industries, like steel, showed strikingly different processes as they declined. The industry was concentrating in the Midwest as it declined nationally; but it was also fragmenting with the emergence of speciality steel producers, fabricators, and distributors. As the fully integrated industry complex decomposed, new opportunities for collaboration around technology, infrastructure, and investment occurred. For example, one of our cases will describe how city government invested in industrial infrastructure in the vicinity of an old steel mill; and the mill's owners, its workers, and a neighboring grass-roots organization supported the city's controversial industrial retention policy.

Firms on the small end of size hierarchies or firms in industries whose market and supplier linkages were being radically redefined more often engaged in cooperative action to advance technology or establish new markets. Similar processes were happening in other Chicago industries—apparel and printing—as well as nationally in motion picture production and automobile manufacturing (Ranney and Wiewel, 1987).

The process of concentration/fragmentation mixed with ownership changes also accounts for many plant closings brought about by the consolidation of plants, disinvestment overseas, or a move to low-wage areas. One of our cases, the Playskool plant closing, addresses just such a closing. Plant closings raise the question of whether the disinvestment calculus ignores local opportunities and skills and, as such, creates opportunities for local challenges to industrial restructuring.

Spatial concentration/dispersal. Geographic restructuring in cities has a number of dimensions, one being the expansion of downtown, its service firms, and related residential enclaves into surrounding industrial and warehouse districts and low-income neighborhoods.

In short, gentrification—the change of neighborhoods to higher-income uses—affects the productive base of cities as well as its residential space. The process pushes firms and industries out, raising the question of where firms go and whether spatial agglomerations for specific industries (e.g., printing) can be recreated (Ranney and Wiewel, 1987). One of our cases recounts the struggle to constrain this industrial displacement. On the other hand, in conjunction with industry fragmentation and specialty niches, the dispersal process also motivated the rehabilitation and subdivision of former single-user factories in neighborhoods where major firms left, thus creating opportunities for neighborhood economic development (Ducharme et al., 1986).

Bifurcation of the labor market. The increasing bimodality of the urban labor market is another dimension of the industry and spatial processes described above. The emerging service sector is characterized by dual concentrations of high-skill, high-wage occupations, such as lawyers and accountants, and low-skill, low-wage occupations such as busboys, chambermaids, and kitchen workers. Plant closings and the contracting out that often accompany restructuring industries reinforce this labor market bifurcation. The resultant real estate pressures have led to extraordinary juxtapositions of land values with high-priced condominium developments abutting public housing projects. Their respective tenants have become competing constituencies for further upscale residential and commercial development, on the one hand, and basic industry, on the other.

CASES

Three cases illustrate how diverse community networks and alternative development ideas may yield more equitable responses to industrial restructuring.[4] The cases are the Playskool plant closing, fighting industrial displacement on Goose Island and in the Clybourn Corridor, and task forces on the steel and apparel industries. Each case study presents the background of the event (including normal public and private responses), ideas and networks that framed the event in a new way or that emerged during the course of the response, the story of what happened, and the outcomes, including replication of approaches and diffusion of development ideas. For each case, we draw attention to what seems most distinctive—whether framing the problem or opportunity, building networks, elaborating development ideas, or impacts. All of the cases are unfolding as of 1988, hence what is offered is a snapshot, one that is also colored by our involvement as participants.

PLAYSKOOL

The city of Chicago's 1984 legal action against Hasbro Industries for closing its Playskool toy factory in Chicago and breaking its Industrial Revenue Bond (IRB) agreement was not the usual municipal response (Giloth and Mier, 1986). As part of a concentrating industry faced with increasing international competition, Hasbro was consolidating its Playskool and Milton Bradley subsidiaries in Springfield, Massachusetts.

Despite extensive plant closings in older industrial cities that have resulted in significant human, neighborhood, and fiscal impacts, few cities have gone beyond offering up more incentives or attempting to soften the hardships of dislocated workers within the straitjacket of the Job Training Partnership Act (JTPA). Nevertheless, there is a growing list of bolder community responses that have used such tools as eminent domain, plant closing ordinances, development agreements, and worker buy-outs (Giloth and Rosenblum, 1987).

Playskool—an old Chicago toy manufacturer of Lincoln Logs—was bought by Milton Bradley in 1968, which in turn, was bought by Hasbro in September 1984. The Playskool closure involved repeated public financial assistance, multiple plant closures and relocations, the rise and fall of electronic games, and corporate mergers.

Public assistance, in particular, may have fueled Milton Bradley's and then Hasbro's plant closure strategy. Employing 700 workers throughout most of the 1970s, Chicago's Playskool plant jumped in employment to about 1200 in 1980, a result brought about by the closure of plants in South Bend, Indiana, and Skokie, Illinois. A city of Chicago loan for new equipment eased the closure of these plants and the transfer of production to Chicago. Finally, Milton Bradley received a $1 million loan from the Massachusetts Industrial Development Authority for the facility to which Playskool productive assets had been transferred.

The Westside Jobs Network (WJN), a group of Chicago community and labor organizations, began to organize one week after Hasbro's September 19, 1984, announcement that it would close the Playskool factory at Christmas. Thus ensued a four-month organizing campaign. The city had funded the Midwest Center for Labor Research (MCLR) and the Center for Urban Economic Development (UICUED) at the University of Illinois at Chicago to establish an early warning demonstration project and community network. The WJN was the product. The WJN included community and development organizations, unions, churches, local public officials, and technical assistance centers.

WJN and its supporters were responsible for the public perception of "community uproar" around the Playskool closure. Images of picket lines, unemployed families, toys, and broken promises soon appeared in newspapers, radio broadcasts, and on coast to coast television. Playskool became the symbol for all that was wrong with the Chicago economy. Its failure to create the 400 jobs promised in exchange for the IRB further aggravated this breach of trust. On December 8, at a large public meeting in a neighborhood church, WJN declared a national boycott of Hasbro toys.

WJN's action inspired other local and national responses to the Playskool closing. Over 100 Playskool workers supported a lawsuit to intervene on the city's case. A local religious coalition filed a friend-of-the-court brief on behalf of Chicago citizens, and economic justice groups from around the country bombarded Hasbro with telegrams and telephone calls.

December 1984 began with Mayor Washington directing city attorneys into court, charging Hasbro with a breach of contract and claiming damages. The court action set a context in which to negotiate a settlement, but formulating a settlement tested the confidence of the community/labor coalition in the city and the unity of the WJN: Member groups were torn between wanting to keep jobs, achieving more equitable severance for the workers, protecting the local industrial community, and assuring that business development would occur in the future.

The settlement reached on January 31, 1985, between the city of Chicago and Hasbro Industries was much less than organizers hoped for, but more than Playskool would have provided in the absence of the community campaign. Hasbro promised to keep the Playskool factory open and retain 100 workers until November 1985, to operate a Playskool Jobs Center that offered $500 to employers that hired ex-Playskool workers, to establish a $50,000 emergency fund for Playskool workers, and, with the city, to find a successor firm or donee for the Playskool facility. Hasbro also agreed to negotiate with new plant users hiring agreements that would give preference to unemployed Playskool workers. These commitments were in addition to the $2.3 million severance package negotiated by union and nonunion employees.

This alternative plant closing response required the participation of community, labor, business, and local government. The city probably would not have taken such relatively aggressive action without pressure from a community/labor coalition. The political risks would have seemed too high. A court suit would not have occurred if Mayor Washington had been insensitive to community concerns and unwilling to consider unconventional actions and support of an early warning network. Continued organizing and winning favorable media and public opinion also forestalled corporate backlash against perceived government meddling.

By no means did the court settlement end the Playskool closing. The settlement required enforcement and the Playskool facility had to be redeveloped. A monitoring of the performance of the Playskool Jobs Center by community organizations showed that few blue-collar workers obtained job placements, particularly female Hispanic workers. Women who eventually found work had average wages of $3.50 an hour, much less than the $5.00- to $8.00-an-hour jobs at Playskool (Kleiman, 1988). On the development side, the Playskool facility is now being redeveloped as a mini-industrial park. The Greater North Pulaski Development Corporation (GNPDC), a community-based development group that had been involved in WJN, is playing a role in the redevelopment. Unfortunately, early hopes to link the redevelopment of the Playskool facility to the employment and retraining needs of dislocated workers have not been met because of developer reluctance.

Beyond the Playskool settlement and aftermath, there were other impacts. Most important, alternative development ideas were reaffirmed, leading to further experimentation with early warning intervention (LaBotz, 1986; Moberg, 1987). In 1985, many of the same community and labor groups initiated the Coalition to Keep Stewart Warner Open. In contrast to organizing after a plant closing announcement, this multifaceted coalition was truly early in its identification of disinvestment in this company that employed 2000. Local managers of another company, Precision Scientific, contacted the city and a neighborhood industrial council because they saw their nonlocal owners disinvesting. By 1987, a local manager buy-out was achieved for this company and 375 jobs retained. A government, union, and company agreement was reached to retain ECKO Products in Chicago—a recent leveraged buy-out of a local firm employing 500. Other, less-concrete successes included two worker buy-out feasibility studies by WJN in 1984-1985 and union, community, and city cooperation to respond in 1986-1987 to the closing LTV: Chicago's last integrated steelworks and an employer of 1000 workers.

The Playskool legal action also added to the groundswell of concern about plant closings and the abuse of public incentives around the United States. In particular, local challenges to closures in Boulder, Duluth, and Louisville have used the Playskool experience. Nevertheless, while Playskool encouraged other creative plant closing responses, federal and Illinois plant closing legislation remained elusive until recently.

INDUSTRIAL DISPLACEMENT

Questioning the costs, benefits, prerogatives, and impacts of downtown development is a risky undertaking. It confronts growth coalition boosterism, challenges the rational voice of the market, and questions the allocation of public subsidies and regulatory exceptions for downtown projects. Although most city governments unquestioningly promote downtown development, a number of examples exist in which downtown dominance has been challenged: linked development agreements and ordinances, stopping big projects, and altering the terms of public financing and ownership of downtown projects (Keating, 1986; Mier, 1984; Shlay and Giloth, 1987).

This case study is about how Chicago is attempting to protect one of its industrial districts from displacement caused by an expanding downtown and upper-income neighborhoods (Schmidt, 1987). Industrial displacement became a public issue because of the research, education, and advocacy efforts of the neighborhood industrial council for the North Branch industrial district (Giloth and Betancur, 1988). The New City YMCA's Local Employment and Economic Development Council (LEEDs) is unusual among industrial councils because of its merging of industrial council activities, such as providing services to area manufacturers with neighborhood development, and advocacy. Moreover, the LEEDs Council is distinguished by its attempt to serve and link the Cabrini Green public housing project and the adjacent North Branch industrial area.

Although business displacement was found to be as frequent as plant closings, its potential consequences were not well understood by the public. Except for community opposition to the upscale redevelopment of an old brewery complex in the late 1970s, normal city of Chicago planning practice was to do what developers wanted. Hence research and public education about industrial displacement and the benefits of balanced urban growth is the theme of this case. In the process of education and advocacy, a loose citywide coalition of community organizations, networks, labor, small businesses, aldermen, and city agencies came together to advocate solutions to industrial displacement. This coalition has been pitted against the developer community and its chief public interest advocate, the Chicago Tribune Company. As a result of this organizing, a number of redevelopment projects have been stopped or redefined, a Planned Manufacturing District (PMD) has passed the city council, and Apparel and Printing Task Forces have made antidisplacement recommendations for downtown-based apparel and printing firms. Several of the negotiated settlements for the contested redevelopment projects have involved financial contributions for industrial land-use planning, commitments for nearby industrial development, and project redesign.

Chicago has not been alone in confronting industrial displacement. In fact, most older industrial cities in the Midwest and Northeast are experiencing similar displacement problems. New York has developed the most ambitious set of policy tools to protect industrial districts, primarily the apparel districts in Manhattan, and promote the industrial retention of relocating firms. Even Baltimore, mecca of downtown boosterism and waterfront development, is experimenting with zoning controls to protect industrial districts (Betancur and McCormick, 1985).

The North Branch manufacturing district is two miles northwest of Chicago's Loop and is part of the North River Industrial Corridor that houses 10% of Chicago's manufacturing jobs, and over 50,000 total jobs. It includes Goose Island and the Clybourn Corridor, both of which are experiencing the piecemeal conversion of industrial property to residential and commercial uses. If speculative redevelopment continues, most firms in these areas are likely to relocate outside of the district due to high rents and property taxes, evictions, lack of adequate space to expand, hard-to-refuse purchase offers, and conflicts with new land uses. Many firms will remain in Chicago, but others may choose to move out of the city altogether (Ducharme et al., 1986).

The city of Chicago and other units of government are pursuing conflicting policies regarding the North Branch district. On one hand, they have supported local manufacturers with capital improvements, financial incentives, support for a local industrial organization, and city plans designating the area as industrial. On the other hand, the city has encouraged redevelopment by granting zoning changes required for residential reuse and by allowing redevelopment proposals to reach near-fruition before public review. Other units of government, such as the Illinois Housing Development Authority, have made loans for upscale housing developments and have been insensitive to the needs of industry as they make riverfront plans and public works decisions.

River lofts. Goose Island is the southern terminus of the North Branch district and is created by the split in the north branch of the Chicago River. Originally a feisty settlement of immigrants, Goose Island became a district of tanners, smelters, and manufacturers at the beginning of this century. In 1988, there were 25 firms and 2,000 jobs on Goose Island along with vacant railroad land and empty factories. Once designated an industrial park under Mayor Jane Byrne in the late 1970s, this plan collapsed when the developer withdrew from the project. Even though many industries remain on Goose Island and others might move there if they were certain of its future, large projects for Goose Island surface with some regularity: the proposed 1992 World's Fair, high-income housing and boat slips, and a sports stadium.

Loft conversions are occurring on Goose Island. The fight over the River Lofts development illustrates the effort needed to convince officials and the public at large that there is industry on the island worth saving. River Lofts is an ambitious project proposed for the southern tip of Goose Island—a $20 million renovation of an old factory for 200 combined work and residence units and commercial uses, including restaurants. Conflict arose because Goose Island industries did not want residential uses on the island, although they did not oppose its commercial development. They reasoned that 24-hour users would eventually object to the noise, the smells, and the truck traffic from nearby industries. River Loft's developers were undaunted: they hired Chicago's premier zoning law firm, mobilized support from other loft developers, and played Chicago politics.

Industries and developers battered city government with letters, meetings, and telephone calls. A public hearing was held and industries and developers met separately with city agencies. Developers offered a compromise that would reduce the number of work and residence units and include lease provisions intended to prevent the harassment of industries. Industries stood firm and told of their plans for $10 million in local investment. After multiple cancellations of the Chicago Plan Commission meetings and widespread press coverage, the alderman for the ward postponed consideration of River Lofts until after his reelection in April 1987.

A compromise agreement was reached in June 1987, only after political operatives communicated to the industries and their industrial council that permanent enemies would be made if their opposition continued. Complete victory for the industries seemed impossible or too costly. Industries did not stop the development but managed to gain future protection of their district.

The final settlement reduced the number of work and residence units and the scope of outdoor restaurants on the site. The developers further agreed to buffer their property from industrial neighbors, establish a mediation system to deal with environmental complaints, and to forgo other types of residential development on the island. The city agreed to investigate whether a cul-de-sac and traffic rerouting on the island could be accomplished to separate more completely River Lofts from its industrial neighbors. Finally, the alderman agreed to support passage of a Planned Manufacturing District ordinance for Goose Island.

Planned manufacturing districts. A question posed early by neighborhood advocates and city planners was whether there were local public policies that could prevent industrial displacement and avoid or set limits on case-by-case zoning fights. One approach, revealed in a survey of 13 cities, was New York's creation of protected manufacturing districts with zoning restrictions and relocation incentives (Betancur and McCormick, 1985). A legislative opportunity to advance such an approach occurred when the alderman of the ward including the Clybourn Corridor, immediately north of Goose Island, declared a moratorium on considering proposed changes from manufacturing zoning until the city developed a workable policy on industrial retention and zoning. He was under pressure from industrial representatives, developers, residents, and numerous city actors every time a proposed loft conversion surfaced.

The design of an industrial protection policy occurred concurrently with such zoning controversies as River Lofts. The Clybourn Corridor, in contrast to Goose Island, has many more loft conversions (including one residential condominium development) and is directly adjacent to the upper-income and ever-expanding Lincoln Park neighborhood. In the course of two years, there were between seven and ten proposed conversions of industrial property in this district for shopping centers, housing, and boutique complexes. Yet the area retained an industrial base of close to 2,000 jobs, including a major specialty steel producer. Also, it had received millions of dollars of public industrial infrastructure in the past decade, particularly roads and sewers. Public hearings on each of these cases and the publication of a research study on the potential negative impacts of industrial displacement provided the rationale for the moratorium.

The outcome of over a year of ordinance drafting, boundary drawing, public hearings, legal strategy, the election of a new alderman, and a mayoral tour was a PMD enabling ordinance and a specific Clybourn PMD application. Chicago's city council held a final set of public hearings in early 1988 and the PMD enabling ordinance passed in the spring of 1988.

The enabling ordinance allows the establishment of PMDs with core manufacturing areas where special and residential uses are limited and buffer zones where a broader range of commercial uses are permissible. Properties in PMDs may be changed to nonmanufacturing only if they are marketed in a good faith manner for manufacturing and new uses do not negatively affect industry.

The lengthy process of ordinance preparation revealed new complexities about industrial displacement. Surprisingly, several businesses opposed the ordinance because they were included within its boundaries; they wanted to get as much as possible for their properties—as their stockholders supposedly demanded. Other firms lobbied to be included in the district. Industries, in general, learned to argue their case in terms of their contribution to Chicago's economy, specifically jobs and taxes. Remnants of the blue-collar residential neighborhood endorsed the PMD concept, but other residential groups were ambivalent: They wanted more residential but reluctantly understood the economic development arguments for preserving industrial districts. A number of citywide groups and several union locals endorsed the PMD concept and became involved in the public hearings. At the penultimate community hearing in March 1987, a majority of the 200 residents and businesses gave their support to PMDs.

Industrial displacement and, more broadly, the goal of balanced urban growth are now a fixture on Chicago's public agenda. Much discussion has ensued about the future of industry in Chicago, the nature and benefits of downtown growth, the development biases evident in regulatory processes such as zoning, and the distribution of public incentives and subsidies. In addition, a network of organizations, including unions, community organizations, and business groups, has emerged to promote alternative development ideas and respond to displacement cases in the neighborhoods. Indeed, cases in many neighborhoods, whether related to downtown expansion or shopping center developers, are being confronted with new confidence. At the same time, downtown growth advocates, developers, and machine politicians are fighting back.

Industrial advocates realize that the long-term industrial displacement challenge is to build industrial districts that are competitive. Industrial districts should have strategic plans to attract new firms, lower inner-city business costs, and help older firms expand and prosper. Compromises will have to be forged that accept light manufacturing and service-type firms, and, as in the case of gaining industrial protection, a partnership of business, the local industrial council, and municipal government will be required. This is already occurring on Goose Island: Three months after the River Loft's settlement, a major developer announced a 1 million square-foot industrial park for an empty 37-acre site.

INDUSTRY TASK FORCES

This case examines the designing of local industrial policies through the creation of task forces, specifically for Chicago's steel and apparel industries. It differs from the previous cases in several respects: Its focus is sectoral—rather than firm-specific or geographic; it involves more formally deliberative problem solving; and the task forces are more narrowly municipal initiatives, although they include broad participation (Wiewel and Siegel, 1987). The task forces changed the perception that manufacturing was dead in Chicago; produced concrete initiatives in the areas of technology, dislocated workers, and business retention; and provided a framework for intervening in plant closings and undertaking legislative lobbying (Alexander et al., 1987).

These industry task forces are indicative of an array of cooperative industrial policymaking that has taken place in Chicago between 1983 and 1987. In most cases, the city of Chicago provided some financial and staff resources. For example, a task force on the printing industry sponsored by UICUED at the University of Illinois at Chicago was completed in July 1987 (Ranney and Wiewel, 1987). The Capital Base Task Force designed and is now implementing the Chicago Capital Fund, an alternative equity capital fund for neighborhood manufacturers (Patterson and Sieros, 1985). The Center For Neighborhood Technology (CNT) has provided technical assistance to small electroplaters attempting to comply with environmental regulations, while advocating for more supportive public policies and cooperative waste disposal techniques (Basler and Kitwana, 1986).

Efforts to design local industrial policies in Chicago confronted opposition ranging from amused disbelief to hostility. There was a common belief that economic development was simply a matter of real estate development, particularly downtown development. Moreover, in 1984, the Commercial Club of Chicago, a gathering of major corporate leaders and the sponsor of Daniel Burnham's 1909 Plan for Chicago, published its strategic plan, *Make No Little Plans*. It made no mention of manufacturing. Even when manufacturing was acknowledged as important, a pervasive feeling existed that it was a national problem that localities could not affect. Changing the terms of the economic development debate in Chicago to focus on jobs and to include manufacturing became a theme of Mayor Washington's *1984 Development Plan* and a function of the industry task forces.

Mayor Harold Washington appointed in 1984 two industry task forces to evaluate the problems of Chicago's steel and apparel industries and to recommend solutions. He was a longtime advocate of manufacturing (his congressional district contained the steel communities). For both steel and apparel, the importance of the remaining jobs and signs of industry resurgence reinforced the need to address the economic and social dislocation resulting from industrial change. Since 1979, approximately 13,000 steelworker jobs were lost in Southeast Chicago and each plant shutdown led to further layoffs by suppliers and distributors. Whereas the steel crisis in Chicago was precipitous, the economic problems of Chicago's apparel industry resulted from long-term decline: in 1983, 365 apparel and textile mill firms employed 10,372 workers in Cook County, down from 29,000 workers in 689 firms in 1964 (City of Chicago, 1986, 1987).

Chicago task forces did bring new people and institutions to the table. The Task Force on Steel and Southeast Chicago had a policy committee of 12 members and working groups that contained an additional 30 persons. Its policy committee included lawyers, academicians, real estate experts, and union and management representatives. Task force staff initially felt that this membership mix was optimal, but in retrospect, representatives of steel-related companies—such as steel service centers and machine tool builders and unions—would have added another important dimension. A source of frustration was the unwillingness of a large, national steel company headquartered in Chicago to participate in the task force. As the task force progressed, however, the company expressed greater interest, eventually meeting with task force members and participating in a steel technology conference. The Apparel and Fashion Task Force, in contrast, included more numerous apparel manufacturers and designers in addition to union representatives and academics. Its policy committee had 20 members, and 23 additional persons were involved in working groups.

The interaction of research and task force deliberations enlivened both task forces. In the steel task force, in particular, the researcher's documentation of a regional Chicago steel industrial complex (i.e., an interdependent network of steel producers, users, and suppliers) and the emergence of steel service centers proved enormously influential in shaping the task force's conclusion that the local steel industry was not dead (Markusen et al., 1985). The apparel task force, in contrast, relied initially more upon the experiences and opinions of industry members, with formal economic and survey research being completed later in order to confirm and polish these insights.

Each task force reflected the problems and characteristics of its industry. The steel task force confronted an industry that was widely regarded as dying by local media and business leadership: integrated steelworks were outmoded. The task force's first step was to overcome this negative perception. With extensive research, they argued that although Chicago's steel sector was undergoing fundamental changes in production methods, distribution patterns, and market position, it remained vital to the local economy. Chicago's steel industry represented a complex of steel users, producers, and distributors that accounted for 250,000 jobs regionally, and steel service sectors, in particular, showed growth. Moreover, as the steel industry rationalized nationally, the more modern, spatially protected, high value-added facilities of the Midwest exhibited relative strength.

The steel task force then identified concrete steps to assist local steel-related manufacturers and the Southeast Chicago community, and sought to involve as many groups as possible in implementation. The task force recommended area development and retraining strategies (City of Chicago, 1986). Moreover, it realized that, even if a mature industry such as steel became a world class competitor again, all the workers who were displaced from that industry would not be reemployed and the positive economic impact of such a diminished industry on nearby communities would not be as extensive as the negative impacts from the earlier closings. Steel task force initiatives included a comprehensive program of technology application and research, an increased business retention effort in Southeast Chicago, actions to ease the impact of projected electricity rate increases, business development and major infrastructure investments, and a mobilization of resources to respond to the needs of dislocated workers. Finally, the task force identified multiple actors to implement both sets of recommendations: local industrial councils; public/private commissions; a regional steel network of unions, churches, and municipalities; and the mayor.

Many recommendations were acted upon during the life of the steel task force and in the past two years. The task force helped to resolve a court suit (of which the city was party) against USX for closing its South Works plant. The result was the temporary saving of 800 jobs and the retention of the site as an industrial park instead of a site for residential or commercial purposes. The task force and the mayor effectively lobbied for the Keyworth Initiative—a federally funded program to advance research into revolutionary steel-making technology—and in February 1987, convened a steel technology conference with Argonne Laboratories and the University of Illinois at Chicago, which brought together steel management, labor, universities, and the public sector. A number of steel-related companies were aided and the city received a major EDA grant in September 1987 to design specialized programs for steel service centers and for dislocated workers. On the drawing boards in early 1988 were a conference on labor and steel technology, and the seeding of a regional steel network.

The Apparel and Fashion Industry Task Force confronted a smaller, less organized industry. Chicago's apparel industry consists of one major menswear producer, a few midsized women's wear firms, and a diversity of small firms. The industry's future strength appeared to be in high value-added, short batch, women's designer clothes. In fact, there were a number of new Chicago designers coming into prominence (Weiss, 1988). Yet, while the steel industry and labor representatives had a clear understanding and common outlook on these issues even before the task force began, bringing the key apparel industry actors together was itself a major step. Compounding the fragmentation within the industry was the highly competitive nature of apparel producers: executives and designers needed to be persuaded that cooperation was important, and that their mutual interest lay in assuring each other's survival.

In response, the task force's recommendations called for the formation of a permanent Apparel Industry Board. Representing the same diverse constituencies as the task force, this body would continue the cooperative research and activities begun during the task force and work with the city and other actors to implement task force recommendations: a business incubator for designers and displaced firms; new technologies that could be shared cooperatively; increased training and retraining opportunities for apparel production workers; and a "Buy Chicago" program that would encourage the consumption of locally made garments (City of Chicago, 1987).

The concrete accomplishments of the Apparel and Fashion Industry Task Force have ranged from convening Fashion Business Seminars for new designers to city lobbying for fair trade for the apparel industry. The city provided first-year funding for the Apparel Industry Board and, with partial funding from companies and unions, in 1988, concluded a preliminary feasibility study for an apparel incubator building that could also house firms displaced by downtown growth.

In spite of concrete initiatives, Chicago's industry task forces have failed initially to ignite a critical level of community, labor, and business commitment. The Economic Development Commission, a public advisory body made up of major Chicago business leaders, has put off setting up a formal subcommittee to oversee the recommendations of the steel task force, in part motivated by the deep-seated belief that manufacturing is a dying sector. Many community groups and labor activists feel that the task forces did not go far enough—for example, the city failed to use eminent domain to landbank the closed Wisconsin Steel site or the closing LTV facility. Some unions and companies felt

TABLE 6.1
Characteristics of Local Industrial Policy Issue

	Playskool	*Industrial Displacement*	*Industry Task Forces*
Origin	Community/labor network	Single-community organization	City initiative
Time frame	Immediate	Immediate/medium term	Long term
Scope	Single firm, single neighborhood	2-3 neighborhoods/ multiple projects	Industry/region
City stake	IRB agreement	Zoning regulation	Planning/incentives
Problem solving	Advocacy, court action, negotiation	Advocacy/research, legislation, negotiation	Planning, lobbying, marketing
Development ideas	Early warning, public trust, local ownership	Balanced-growth manufacturing	Manufacturing industrial complex
Mode of collaboration	Dynamic, multiple centers, mobilization	Dynamic, multiple centers, mobilization	Formalistic and institutional, city-centered, missing actors
Leadership	Strong, multiple, diffused	Weak, multiple, diffused	Weak, government, institutional
Allocation of costs/benefits	More worker and community benefits	Protected industries and jobs, developer contributions	Promotion of manufacturing
Challenges	Network building, public dialogue	Framing the problem, public dialogue	Framing the problem, public dialogue, implementation
Outcome	Settlement, redevelopment	PMD ordinance, redesigned projects, manufacturing development	Many specific initiatives and responses
Problems	City/network friction, intracity friction	Developer/*Tribune* opposition, intracity consensus-building	Lack of city/public mobilization

that the city did not lobby hard enough for trade protection and was unwilling to close down apparel sweatshops. Surprisingly, some progressive planners even complained that the city had focused too much on dying industries instead of growing sectors such as printing. Finally, other city departments have not been inspired to revamp the dislocated workers program or to initiate land-use planning for Southeast Chicago.

CONCLUSION

We have tried to examine how local development politics in Chicago became a context for a reaction and response to economic restructuring. Table 6.1 summarizes the three case studies, with particular attention focused on issue characteristics. While it is largely self-explanatory, several points, taken together, highlight the political dimension of Chicago's efforts at economic development innovation. Specifically, we are concerned with the elements that emphasize network building: the role of leadership, communication within networks, and the potential for political mobilization (Mier and Giloth, forthcoming).

The Playskool and industrial displacement cases originated in community advocacy. Beginning in a web of diverse organizations, the process of network building was easier, and the potential for the emergence of multiple leaders greater. Because the closing of the Playskool plant was a real threat, while industrial displacement was more long term (at least in Clybourn/Goose Island), the Playskool mobilization was not as difficult but not as lasting. Opposition to the closing was facilitated by the formulation of colorful metaphors and organizing language (local ownership, early warning, breach of trust). An industrial displacement strategy could build on the same language (manufacturing, blue collar), enrich it and link it with broader development metaphors (balanced growth), and broaden the network of participants (more grass-roots, labor, and citywide participation). Each case had multiple leaders, multiple centers of impetus, and broader support than the immediate neighborhood in which the major events were occurring.

A major impediment to this networking, however, is the challenge of social integration across racial boundaries. In their own ways, each case illustrates the difficulty of building multiracial alliances: it is easiest in the most concrete situation, i.e., the one linking black and Hispanic workers, their neighborhoods, unions, and local groups—the Playskool plant closing. With industrial displacement, ties among black public housing tenants, Hispanic neighborhoods to the west, small businesses, and labor slowly emerged during the years of organizing. Yet these experiences did expand the language of development across racial boundaries such that metaphors like balanced growth and linked development had common meaning for different groups.

The industry task forces were, in important respects, the most abstract and politically elusive, and thus were the most difficult to evaluate. They were not the product of an immediate crisis, although each derived from decade-long sectoral disinvestment. As such, the task forces did not conduct their work in a mobilization atmosphere. To this extent, they more easily involved those formal organizations with planning and research orientations—units of government, universities, and business and civic organizations that routinely participated in civic study efforts. The greatest difficulty for these task forces was framing their problems in such a way as to build a constituency and articulate a distinctive and viable local government role. Nonetheless, they enriched public awareness that development opportunities remained even in shrinking economic sectors.

Taken together, all three cases focused local political debate on industrial development. They testified for a development policy that was fair, redistributive, open, and participatory. Broad-based constituencies supported forceful political action, such as the suit against Playskool's parent corporation, Hasbro Industries. In the five-person aldermanic race in the area of the Clybourn Corridor, the two leading candidates made industrial preservation central to their campaign planks. Throughout his own reelection campaign, Mayor Washington maintained a focus on industrial retention and on a nuts and bolts approach to industrial development, even in the face of strong criticism from the *Chicago Tribune*, the city's largest newspaper. Finally, before his untimely death in November 1987, Mayor Washington planned to inject the basic philosophy and concrete approach of his development program into national politics, working first through the U.S. Conference of Mayors and then through the Democratic party platform shaping process. He was confident that he could promise broad-based popular support for an agenda emphasizing people, jobs, and industry.

How does local experimentation connect and accumulate to become effective nationally? How do alternative ideas and networks expand beyond local boundaries without becoming formalistic or diluted in meaning? More specifically, are Chicago's circumstances unique, especially given the extraordinary political vision and charisma of Harold Washington, the extenuated period of political conflict and mobilization (i.e., Council Wars) between 1983 and 1986, and Chicago's legacy of machine politics and racial segregation?

A number of recent works have begun to map the terrain of alternative economic development, although they have not directly focused on the diffusion issue (Bruyn and Meehan, 1987; Clavel, 1986). Distinctive responses to economic restructuring are occurring across the United States under very different political circumstances. Moreover, there is a growing infrastructure of technical assistance centers, organizational networks, and foundations that serves community economic development—one that may both expand networks and control local experimentation (Pierce and Steinbach, 1987).

Building local to national political connections is already occurring through professional associations and political structures such as the Democratic party and the U.S. Conference of Mayors. These routes are important but limited. Recent sociological thinking about community decision making underscores the importance of interorganizational linkages or social networks that connect resource-mobilizing individuals (Bellah et al., 1985; Brown and Detterman, 1987; Wiewel and Hunter, 1985). By exploring feminist consciousness-raising, working conditions, ethical values, and family life, social psychologists are increasingly examining the processes by which small support groups begin to network regionally and nationally (Gilligan, 1982; Rubin, 1985). These perspectives on expanding local connections complement and nourish more empirical work looking at a variety of modern populist social movements (Boyte and Riessman, 1986; Boyte et al., 1986).

Local experimentation that responds to economic restructuring in more equitable ways faces additional challenges. First, the downtown/neighborhood metaphor, while powerful, is limited. It tends to generate either/or trade-offs and muddled thinking about economic growth, development, and equity. Second, collaborative efforts among community, labor, and government often leave a residual of ill will that comes with compromise, thus giving incentive for urban actors to pursue more solitary advocacy. Moreover, no adequate body of organizing, public administration, or leadership theory exists to guide these collaborations through the thicket of community conflict (Mier, 1987). Fourth, experimental projects require long gestation and implementation periods that often exceed community political cycles and media attention spans. Fifth, experimentation that challenges economic restructuring eventually evokes opposition from the "powers that be," which in turn tests collaborative efforts even more, thereby aggravating all of the above. Finally, local governments have limits; yet experimental action raises expectations.

NOTES

1. Our framework for understanding Chicago's experiments with economic development and the role of networks, ideas, and metaphors derives from Clavel (1986), Evans and Boyte (1985), Schon (1983), and Whyte (1982). Our thinking on how social inventions diffuse has been influenced by Goodwyn (1978) and Schon (1971).

2. Equity planning articulates the primacy of specific populations in need (e.g., the poor and disenfranchised) in planning goal statements, policy plans, or reformed government processes (e.g., zoning, capital budgeting).

3. Organizing training centers included the National Training and Information Center, Midwest Academy, Associated Colleges of the Midwest Urban Studies Program, and the Urban Planning Program at the University of Illinois at Chicago. Formal networks of neighborhood organizations included the Chicago Rehab Network, the Center for Neighborhood Technology, and the Chicago Association of Neighborhood Development Organizations (CANDO). A major network of political activists was the Task Force on Black Political Empowerment. The Center for Urban Economic Development (UICUED) at the University of Illinois at Chicago, established in 1978, served as a resource base for many community organizations.

4. Our case studies draw extensively from articles and reports written with associates with whom we collaborated during this period: Stephen Alexander, John J. Betancur, Donna Ducharme, Margie Gonwa, Joshua Lerner, Lynn McCormick, Kari Moe, Raffaella Nanetti, Irene Sherr, and Howard Stanback.

REFERENCES

ALEXANDER, S., R. GILOTH, and J. LERNER (1987) "Chicago's industry task forces: joint problem-solving for local economic development." Economic Development Quarterly 1 (4, November): 352-357.

BARRET, P. (1983) The Automobile and the City. Philadelphia: Temple University Press.

BASLER, S. and M. KITWANA (1986) The Impact of Environmental Regulations on Small Manufacturers in Chicago. Chicago: City of Chicago.

BELLAH, R. N., R. MADSEN, W. M. SULLIVAN, A. SWIDLER, and S. M. TIPTON (1985) Habits of the Heart: Individualism and Commitment in American Life. Berkeley: University of California Press.

BENNETT, L., G. D. SQUIRES, K. McCOURT, and P. NYDEN (1987) "Challenging Chicago's growth machine: a preliminary report on the Washington administration." International Journal of Urban and Regional Research (Fall): 351-362.

BETANCUR, J. J. and L. McCORMICK (1985) Industrial Displacement in Major Cities and Related Policy Options. Chicago: University of Illinois, Center for Urban Economic Development.

BOYTE, H. C., H. BOOTH, and S. MAX (1986) Citizen Action and the New American Populism. Philadelphia: Temple University Press.

BOYTE, H. C. and F. RIESSMAN (1986) The New Populism: Politics of Empowerment. Philadelphia: Temple University Press.

BRANDY, D. (1987) "Local development policy in the 1980's." Journal of Planning Literature 2 (2, Spring): 136-152.

BROWN, L. D. and L. B. DETTERMAN (1987) "Small interventions for large problems: reshaping urban leadership networks." Journal of Applied Behavioral Science 23 (2, Fall): 151-168.

BRUYN, S. and J. MEEHAN (1987) Beyond the Market and the State: New Directions in Community Development. Philadelphia: Temple University Press.

City of Chicago (1986) Building on the Basics: The Final Report of the Mayor's Task Force on Steel and Southeast Chicago. Chicago: Author.

City of Chicago (1987) Cooperation for Survival and Growth: New Designs for Apparel Manufacturing in Chicago. Chicago: Author.

CLAVEL, P. (1986) Progressive Cities. New Brunswick, NJ: Rutgers University Press.

DUCHARME, D., R. GILOTH, and L. McCORMICK (1986) Business Loss of Balanced Growth: Industrial Displacement in Chicago. Chicago: City of Chicago, Department of Economic Development.

EVANS, S. and H. BOYTE (1985) Free Spaces. New York: Basic Books.

FAINSTEIN, N. I. and S. S. FAINSTEIN (1987) "Economic restructuring and the politics of land use planning in New York City." Journal of the American Planning Association 53 (2, Spring): 237-248.

GILLIGAN, C. (1982) In a Different Voice. Cambridge, MA: Harvard University Press.

GILOTH, R. (1985) "Organizing for neighborhood development." Social Policy 15 (3, Winter): 37-42.

GILOTH, R. and J. BETANCUR (1988) "Where downtown meets neighborhood: industrial displacement in Chicago: 1978-87." Journal of the American Planning Association 54 (3, Summer): 279-290.

GILOTH, R. and R. MIER (1986) "Democratic populism: the case of Playskool." Cities: An International Quarterly of Urban Policy 3 (1, February): 72-74.

GILOTH, R. and S. ROSENBLUM (1987) "How to fight plant closings." Social Policy 17 (3, Winter): 20-26.

GOODWYN, L. (1987) The Populist Moment: A Short History of the Agrarian Revolt in America. New York: Oxford University Press.

GREENHOUSE, S. (1986) "Chicago does its job hunting at home." New York Times (November 30).

GRIMSHAW, W. (1984) "Is Chicago ready for reform? or, a new agenda for Harold Washington," pp. 141-166 in M. G. Holli and P. M. Green (eds.) The Making of the Mayor: Chicago 1983. Grand Rapids, MI: Eerdman.

"Help for a day or even an hour." (1987) New York Times (May 24).

JACKSON, J. L. (1987) Straight from the Heart. Philadelphia: Fortress Press.

JUDD, D. R. and R. L. READY (1986) "Entrepreneurial cities and the new policies of economic development," pp. 209-248 in G. E. Peterson and C. W. Lewis (eds.) Reagan and the Cities. Washington, DC: Urban Institute Press.

KEATING, D. (1986) "Downtown development to broader community goals." Journal of American Planning Association 52 (2, Spring): 133-141.

KLEIMAN, C. (1988) "Stereotypes plaguing blue-collar women, too." Chicago Tribune (February 29).

KRUMHOLZ, N., P. COSTIGAN, and D. KEATING (1985) "Chicago works together: 1984 development plan"[Book Review]. Journal of the American Planning Association 51 (3, Summer): 395-396.

LaBOTZ, D. (1986) "How can communities fight plant shutdowns." Neighborhood Works 9 (1, January/February): 9-11.

LERNER, M. (1988) "The legacy of the sixties for the politics of the nineties." Tikkun: A Bimonthly Jewish Critique of Politics, Culture, and Society 3 (1, January/February): 44 ff.

MARKUSEN, A. with J. LERNER, W. PATTON, J. ROSS, and J. SCHNEIDER (1985) Steel and Southeast Chicago: Reasons and Opportunities for Industrial Renewal: A Research Report to the Mayor's Task Force on Steel and Southeast Chicago. Evanston: Northwestern University, Center for Urban Affairs and Policy Research.

McKNIGHT, J. and J. KRETZMANN (1984) "Community organizing in the 80s: toward a post-Alinsky agenda." Social Policy (Winter): 15-17.

MIER, R. (1984) "Job generation as a road to recovery," pp. 160-172 in P. R. Porter and D. Sweet (eds.) Rebuilding America's Cities: Roads to Recovery. New Brunswick, NJ: Center for Urban Policy Research.

MIER, R. (1987) "A note on planning's future." Colloqui: Journal of Planning and Urban Issues (Spring): 52-54.

MIER, R. and S. GELZER (1982) "State enterprise zones: the new frontier. Urban Affairs Quarterly 18 (1, September): 39-52.

MIER, R. and R. GILOTH (forthcoming) "Civic leadership: community problem solving for tomorrow," in S. Lourenco and J. Smilansky (eds.) Handbook for Leadership Development. San Francisco: Jossey-Bass.

MIER, R., K. MOE, and I. SHERR (1986) "Strategic planning and the pursuit of reform, economic development and equity." Journal of the American Planning Association 52 (2, Summer): 277-289.

MOBERG, D. (1987) "Back from the brink." Chicago (August): 99-101, 123.

PATTERSON, J. and C. SIEROS (1985) Plan of Action: Community Equity Corporation of Chicago. Chicago: City of Chicago, Department of Economic Development.

PIERCE, N. and C. STEINBACH (1987) Corrective Capitalism. New York: Ford Foundation.

RANNEY, D. and W. WIEWEL (1987) The Graphic Communications Industry in the Chicago Metropolitan Area. Chicago: Center for Urban Economic Development, University of Illinois.

RUBIN, L. (1985) Just Friends. New York: Harper & Row.

SCHON, D. A. (1971) Beyond the Stable State. New York: Random House.

SCHON, D. A. (1983) The Reflective Practitioner. New York: Basic Books.

SCHMIDT, W. E. (1987) "Chicago plan seeks to prevent loss of factories." New York Times (December 10).

SHLAY, A. B. and R. GILOTH (1987) "Social organization of a land-based elite: promoters of the 1992 World's Fair." Journal of Urban Affairs 9 (Winter): 305-324.

SQUIRES, G., L. BENNET, K. McCOURT, and P. NYDEN (1987) Chicago: Race, Class, and the Response to Urban Decline in Chicago. Philadelphia: Temple University Press.

STANBACK, H. and R. MIER (1986-1987) "Economic development for whom? The Chicago model." Review of Law and Social Change 15 (1): 11-22.

WEISS, D. (1988) "The Chicago look." Today's Chicago Woman (March): 11.

WHYTE, W. F. (1982) "Social inventions for solving human problems." American Sociological Review 47 (February): 1-13.

WIEWEL, W. and A. HUNTER (1985) "The interorganizational network as a resource: a comparative case study on the organizational genesis." Administrative Science Quarterly 30: 482-496.

WIEWEL, W. and W. SIEGEL (1987) "Industry task forces as pragmatic planning: the effect of ideology, sponsorship, and economic context on strategy selection." Presented at the annual conference of the American Collegiate Schools of Planning, Los Angeles.

7

Space, Time, and Economic Restructuring

ROBERT A. BEAUREGARD

ANY UNDERSTANDING of the consequences of and resistance to economic restructuring cannot be developed independently from a sense of the way time and space fragment the phenomena and affect our interpretations. Each poses fundamental dilemmas to the social theorist. As studies are conducted and theories disseminated, the reality on which they are based changes. The theorist, despite disciplinary inclinations, automatically becomes a historian, while well-tuned explanations rapidly decay. What we knew about economic restructuring in 1978 seems terribly dated in 1988. Simultaneously, the extension of events and behaviors, and the diffusion of their consequences across space make analytical focus a relentless problem. Just as we begin to understand economic restructuring, it shifts location or expands in spatial scale.

Time and space, of course, are not insurmountable problems. Explanations do manage to have temporal range, and limits. Many do this by utilizing the sluggishness of structure as the plot of the analysis. Others rely on the chronological limits of agency to bracket their investigations. Theorists overcome the boundary problem, too. Events and issues can be spatially delineated precisely because diffusion involves friction and actions have limited spatial reach. At some point, explanation is no longer served by widening the spatial focus. In our everyday conduct, we learn to take time into account and to live in a world of daily rhythms and long historical trends. Our lives unfold within places, and our influence spreads spatially and then diminishes. Only when we fail to acknowledge adequately the weight of time and space do we find ourselves sending forth faulty explanations and acting inappropriately.

The large and expanding literature on economic restructuring struggles with these issues. Economic restructuring is inherently a temporal and spatial notion. In fact, many observers reference the spatial nature of economic restructuring (e.g., Markusen, 1985; Massey, 1984; Massey and Meegan, 1982), while almost all provide approximate historical delineations. Fewer theorists, however, address its nature and consequences along both dimensions simultaneously. At the same time, specific political responses to restructuring are widely recognized but seldom integrated with an understanding of how economic restructuring serves as both an impetus and also as a constitutive element in the content and strategy of those responses.

The purpose of this chapter is to explore the concurrent temporal and spatial nature of economic restructuring and political resistance during the postwar period in the United States. My thesis is not simply that time and space are important and interrelated, but that space shatters time by distorting the simultaneity of structural events. More precisely, the uneven spatial development of capitalism refracts the ongoing economic restructuring in ways that dampen its overall impact and weaken political resistance by undermining common consciousness and by blocking the channels of mass mobilization and collective political action. Political resistance to restructuring is fragmented by uneven spatial development.

This theme reflects the structure of the chapter. Two broad issues are addressed. The first is the way in which uneven spatial development causes economic restructuring to occur at different times in different places, thus making the timing of structural change elusive. Primary emphasis is given to how we think about structure; structure is the key to many theoretical problems. The second issue concerns the extent to which uneven development mediates spatially and temporally the political response to economic restructuring. Here we look at how capitalism's unevenness undermines class consciousness, stifles mobilization, and forestalls effective political strategies. The result is a reactionary and individualistic conservatism, rather than a collective and unified progressive response on the part of labor.

RESTRUCTURING IN SPACE AND TIME

Economic restructuring is a complex term whose meaning is not always clearly articulated. Specification requires that it be positioned along three dimensions: one functional, a second temporal, and a third structural. In fact, two terms exist. One refers to transformations in the mix of goods and services being produced within an economy and is frequently measured by the relative shares of employment across industrial sectors. This is industrial restructuring, an approach found often within case studies on postwar urban development (Beauregard, forthcoming-a; Fainstein et al., 1986). The other refers to a broader set of changes in the nature of the economy, not only in its products and the distribution of employment but also in the social relations of production (e.g., extent of unionization, part-time versus full-time jobs), the means of production (e.g., robotics, handcrafting), and the forces of production (e.g., profit requirements, market demand). As Glickman (1987: 81) writes: "Structural change (or economic restructuring) involves long-term shifts in the composition of demand, production and occupational patterns; new technology; a changing international division of labor; shifts in relative prices; and evolving location patterns (both migration and industrial spatial restructuring)." *Economic restructuring* thus subsumes *industrial restructuring*, though many times, the terms are used interchangeably.[1] Any restructuring of the mix of industries in an economy is tightly linked, in practice and theory, to more profound changes in other economic realms.[2]

Both terms contain a critical temporal dimension. Fainstein and Fainstein (Chapter 1 of this volume) put the issue bluntly: Is economic restructuring a discontinuity in the history of capitalism or simply a gradual transition only recently appreciated? The key words here are *discontinuity* and *gradualism*. The first references a "break" in chronological time such that one can, simply put, specify the date at which capitalism went from one structural form to another. For example, Storper and Scott (1986: 3) claim a new historical reality for capitalism: "Since the second half of the 1970s, capitalism in America, Western Europe, and the world at large seems to have embarked on a radically new course in comparison with the economic and political structures that were set in place in the decades immediately following World War II."

By contrast, the second position is highly evolutionary: no sudden shift but a slow transition by which incremental adjustments accumulate to produce a structure in motion. Gradualism disallows the specification of a discrete date. Instead, the theorist, if attending to history at all, speaks in terms of vast time periods or "ages" whose temporal brackets are centuries. The change from mercantilism to capitalism occurs not at once but slowly and incrementally, perceptible only in retrospect. Discontinuities, however, are the essence of history, gradualism the realm of dynamics (Beauregard, 1988), yet the two are not mutually exclusive. Time can take many forms simultaneously. Our lives are biologically paced, our world structured by the rhythms of work and home, and our fate embedded in deep historical tides. Thus the choice is not simply binary. Numerous positions exist along the path of time's arrow.

Nonetheless, any discussion of time is difficult without reference to the third dimension: structure. Economic and industrial restructuring are profoundly concerned with the basic and enduring relationships that give to economies and social formations their essential qualities. Structure means sluggishness. Structure sets the social formation into a longer timeframe than that conferred by agency and contingency. Structural changes are sea changes, and often revolutionary: "*Restructuring* is meant to convey a break in secular trends and a shift towards a significantly different order and configuration of social, economic and political life" (Soja, 1987: 178). Agency and contingency, the antithesis of the structural thesis, imply rapidity and instill a pervasive instantaneity and continuity to social change. Incremental adjustments replace qualitative and profound discontinuities.

Yet, it would be foolish to do otherwise than to theorize all social formations as in a state of flux not only in their contingent relations but in the structure itself. Structure can be both enduring and in motion, as the structurationists (see Moos and Dear, 1986) have so accurately argued. In addition, "structure is both medium and outcome of social practices" (Giddens, 1979: 5). To this extent, restructuring is constant if by restructuring we mean the reproduction of a structure in flux.[3]

On the other hand, deep and sudden structural change, almost by definition, is not easily arranged. A combination of commitments to enduring patterns of relationships and the ontological security that they confer establishes the stability of structures. Large-scale social movements or basic contradictions leading inexorably to crisis are the accepted forces of structural change, not the constant and small-scale adjustments involved in the reproduction of the structure. While social movements and unyielding contradictions are difficult to produce, neither are they merely random conjunctures that appear without warning. The conditions for their development must be established and the moment seized. That crises are more often predicted than experienced speaks to the resilience of structure, to the limits of theory, and to the pervasiveness of ideology. Last, incremental nonstructural changes might well occur simultaneously with crises, and be entirely independent of them. The social formation thus experiences in tandem structural change—restructuring—and nonstructural adjustments.

Those whose work articulates qualitative discontinuities in capitalism often link their sense of economic restructuring with the notion of crisis. In part, such a position is indicative of a Left, particularly Marxist, perspective that understands capitalism in terms of internal contradictions that automatically produce crises and that eventually lead to capitalism's demise. Castells (1980), for example, pinpoints the 1970s in the United States as such a period, one that followed upon a long-term postwar economic boom. Since that time, American capitalism has been undergoing a structural crisis, a situation in which "it becomes impossible to expand or reproduce the system without a transformation or reorganization of the basic characteristics of production, distribution, and management, and their expression in terms of social organization" (Castells, 1980: 8). Bradbury (1985: 39) agrees, and states this clearly in his definition of restructuring: "An embracing term coined to describe the empirical and theoretical experience in a mode of production during a crisis phase." Moreover, he views this restructuring, and subsequent and related crises, as a "logical consequence of the path of capitalist development" (p. 38).

In other words, restructuring and crises are not only functionally linked but intrinsic to the capitalist social formation. Still, Bradbury does not claim restructuring as a permanent condition of capitalism. Restructuring and crisis are neither continuous nor periodic phenomena. Rather, capitalist economies are characterized by two different time paths: periodic cycles and restructuring phases that appear only intermittently but inevitably.

From a Marxist perspective, however, structure (and thus restructuring) is both a comfortable (and comforting) and a disturbing notion. Structure has a long history in Marxist thought, being the essence of the base-superstructure metaphor (Williams, 1977: 75-82), which has proved so problematic for all but Marxist economic determinists. The economy—the base or structure—is the source for the qualities of all other dimensions of the social formation, not just the social relations of production but also the cultural, ideological, political, and social relations outside the base; that is, in the superstructure. In short, the superstructure sits upon and is supported by the base. The structure, in an architectonic sense, extends throughout the theoretical edifice. Without it, the edifice would not and could not stand. To this extent, restructuring involves qualitative transformations in the base, which then reverberate, hardly dampened, throughout society. Changes outside the economy, then, occur only after and are directly related to changes in the sphere of production.[4] While most Marxist theorists have abandoned the mechanistic and spatial logic of the base-superstructure metaphor, recidivism often strikes. This explains, in part, the attraction of economic restructuring as a concept.

Without falling into economic determinism, Fainstein and Fainstein (Chapter 1) utilize a variant of the base-superstructure distinction in thinking through the temporal dimension of economic restructuring. Basically, they view the social formation as itself containing divisions—urbanism, the household, social relations of production, and the state—whose boundaries encapsulate relatively homogeneous structure and behavior. The social formation is not of one cloth. Implicit is that each division or sphere has a certain degree of autonomy, not only from the economic base but also from each other, and different forces shape and define each of them. Thus their concern extends beyond the existence of a qualitative discontinuity in a social formation. More precisely, they are interested in the parallel restructuring of various spheres as well as the discontinuities within them.

The Fainsteins conclude that many of the changes that we are witnessing, for example, a redefinition of the working class and changes in the pattern of urban development, are variations on long-standing trends, whereas in the sphere of reproduction, the family has undergone fundamental transformation along with the role of women in society. They acknowledge only two significant changes in world capitalism: the shifting of the balance among capitalist nation-states to the Pacific basin and the integration of Third World peripheries into the world capitalist economy. Their analysis suggests that we must think in terms of the structures of the social formation, pay attention to the mechanisms that transmit forces from one structure to another, and search for correspondences (and the relations that bind them) among spheres.

From another perspective, structure is a disturbance in the consistency of Marxist social theory. The inherent crisis tendencies of capitalism are not just linked to its contradictions, but also to its anarchic quality. Marxist and other Left critics frequently view capitalism as producing chaos rather than integration, incongruities rather than coherence, and inefficiencies. The anarchy of capitalism is used, in fact, as one of the pillars of arguments for the rise of the capitalist state, whose functions include protecting the long-run and broad-based interests of capitalism from being eroded by intracapitalist competition and by the lack of capitalists' attention to the social consequences of their efforts (see, for example, Kay, 1979: 112-116).[5] Boyer (1983) uses this as a central theme in her analysis of the rise of city planning in the United States. Industrial capitalism not only ignored the numerous public problems that it created, but also generated a hodgepodge of land uses lacking a rationalized transportation network that would allow the city to function as an efficient part of the productive apparatus. City planning emerged to develop the city "as a perfectly disciplined spatial order" (Boyer, 1983: 60) serving the needs of capital accumulation.

Anarchy and structure, however, are not easily reconciled. Anarchy suggests initially either a lack of structure, that is, a randomness to the economy and social formation, or an oxymoron: anarchic structure. Certainly, reconciliation is not impossible. Rather, the task is too often overlooked by those who hold onto a bundle of disparate concepts: structure, anarchy, and intermittent crises.[6] Restructuring incorporates—or could be made to—these various notions, and the resultant sense of structure would enable us to reflect on the perseverance and change of social formations, and thereby sharpen our understanding of time and space in the restructuring of capitalist society.

TIME

Structure is important in giving meaning to time. Out of the interaction of structural change and a sensitivity to history emerges the penchant for periodization, a theoretical device central to economic restructuring. It involves a division of time into historical clumps or periods of often unequal chronological span. As Jameson (1984: 16) claims, in a less neutral manner, periodization tends to "obliterate difference, and to project an idea of the historical period as massive homogeneity (bounded on either side by inexplicable 'chronological' metamorphoses and punctuation marks)." The intent is to organize structure in a way that has temporal meaning in terms of qualitative transformations. Thus, as an example, the mid-eighteenth century to the early decades of the nineteenth century become the Age of Mercantilism, from then to the late nineteenth century the Age of Laissez Nous Faire, and afterward the Age of Corporate Capitalism (Williams, 1973). Even the very common notion of "postwar" is intrinsically periodizing—World War II as a watershed qualitatively separating one era from another. Periodization thus incorporates both discontinuities and gradual transformations. History proceeds through incremental changes interrupted at uneven intervals by sharp disjunctures in the structural underpinnings of society.

Theorists interested in industrial and economic restructuring have turned frequently to this theoretical tool in their quest to understand present transformations. First, they have accepted the postwar period as a time of conceptual and empirical integrity. The 1920s and the 1930s, significant years for comprehending changes that occurred after World War II, are all but forgotten.[7] The premise, often left implicit, is that the Depression and World War II were qualitative discontinuities, wholly restructuring society, and thus establishing the baseline for the postwar focus. Within the postwar period, a brief recession was followed by a long period of prosperity, labor peace, international hegemony, and expanding affluence, all rooted in the national and international transformations engendered by World War II. Ultimately, growth and harmony were interrupted by the deep recession of 1973-1975 with its stagflation, numerous plant closings, rising federal deficit, and the onset of debtor status for the U.S. economy. The seemingly sudden deflation of the economic boom gave birth to the notion of economy restructuring. Thus such observers as Bluestone and Harrison (1987) periodize the postwar period as follows: 1963 to 1973 were a time of prosperity; 1973 to 1979, years of crisis; and 1979 to 1985, a period of recovery. The wild fluctuations in the stock market in late 1987 will likely extend and end the period of recovery and serve as the division that generates a fourth period.

Even those theorists who deemphasize crises and eschew perspectives that utilize historical discontinuities recognize the temporal unevenness of capital's history and ultimately, or implicitly, adopt a structural periodization. Clark et al. (1986), for example, argue that accurate insights into the U.S. economy can only be built upon theories that view economic change in terms of spatial and temporal adjustments. These adjustments are incremental, multitudinous, reversible, and pervasive. Capitalism functions through a never-ending and vast series of adaptations between and among workers and establishments, and involving conditions of employment, prices, and even migration. Sharp discontinuities in regional structure are explicitly rejected, and a tendency to crisis is not part of the theory.

On the other hand, regional adjustment theory is based upon a spatial division of labor perspective and an explicit and central role for time and space in creating and mediating the response to economic uncertainty. The authors' goal is "a genuinely dynamic theory of regional adjustment: to theorize the temporal structure of plans and the spatial consequences of attempts to realize them" (Clark et al., 1986: 17). The objective is to resist neoclassical theory's annihilation of time and space. Yet, within their empirical analysis, a bracketing, even periodization, seems to be in force. The quantitative arguments begin in the mid-1950s and end in the mid-1970s. That time span corresponds, more or less, to the period of postwar prosperity within which few economic theorists see anything but normalcy and continuity. During that time, capitalism's structure remained relatively stable and intact. Thus an approach to restructuring that rejects sharp discontinuities and crises, and that makes time a central moment in the theory, implicitly adopts a structural perspective on temporal change (Beauregard, 1988).

In contrast, an acceptance of contradictions, crises, and anarchy does not lead inexorably to the position that contemporary restructuring represents a significant alteration of capitalism. Even though Chase-Dunn (1984) recognizes that the growth in the size and importance of transnational firms, the further integration of the world-economy, deindustrialization in the core and industrialization in the periphery, and the increased proletarianization of the world labor force have sped up in the postwar period; he still argues that the deep structural logic of capitalism has not changed. All of the constant structural features remain: economic and military competition among nation-states, the core-periphery dualism of labor, the differentiation of labor along a continuum from protected to coerced, and commodity production for world markets. World capitalism is undergoing "rapid changes of scale and apparently new institutional forms" (Chase-Dunn, 1984: 75) and is driven by an "anarchy of production decisions" (p. 85), but these are continuations of long-term processes and not indicators of a fundamental alteration of the system's logic. Thus restructuring is possible under a variety of theoretical assumptions about the dynamics of capitalism. Regardless of whether one characterizes it as a qualitative discontinuity with the past or a gloss on capitalism's underlying logic, temporal analysis remains essential. Though restructuring might be reduced to a multitude of adjustments between and among labor and capital, periodization cannot be avoided, even if the treatment of time remains problematic.

SPACE

Our indecision about the temporal patterning of economic restructuring is exacerbated by uneven spatial development. The temporal uneven development of capitalism manifests itself in an uneven spatial pattern, whereas uneven spatial development is inherently a temporal phenomenon: it involves the *movement* of development across places (Smith, 1984), a movement that takes place in time and not in the instantaneous world of the neoclassical economic theory (see Clark et al., 1986: 14-16).

The interaction is clear in Clark's (Chapter 2) study of the Pittsburgh region. At the height of industrial capitalism, the whole region boomed with steel production both within the city and along the river valleys. Corporate activities and administrative labor were concentrated in the city, steelworkers in the mill towns. Core and periphery were highly integrated. With the restructuring of the economy in the late 1970s, the steel industry went into precipitous decline and the consequences fell unevenly across the region. At the present time, the mill towns have lost their economic base and exist in the throes of depression. Pittsburgh, instead of falling into decline, has experienced an emergence of an administrative/health/high-tech/university complex, which has maintained the economic viability of the city. In a matter of two decades, the landscape of value has changed to one where an affluent core, once dependent on its productive periphery, has now become an isolated economy within a region of decline. The restructuring not only severed the ties between the city and its hinterland, but also integrated the city into another industrial complex (Scott, 1982). Thus, while Pittsburgh's restructuring represents a temporal shift, it is also profoundly spatial.

For most theorists, restructuring is a spatial phenomenon. Bradbury (1985), for example, views the restructuring of the industrial economy as a geographic event. The economy is a space economy whose workings involve linkages across places and whose consequences reverberate throughout those places, raising (albeit temporarily) certain regions along the growth curve and relegating other regions (again temporarily) to the path of decline. The best-known example in the United States is the decline of the manufacturing-based Northeast and Midwest during the height of economic restructuring in the mid-1970s. Formerly prosperous regions driven by the engine of industrial capitalism, these regions went into an economic tailspin with the industrialization of peripheral countries, the relocation of manufacturing operations to the South, and the erosion of steel and automobile production by foreign competition. The South experienced unprecedented growth. By the mid-1980s, however, the Sunbelt was beginning to stagnate and the regional economy of the Northeast was flourishing. Regional restructuring, then, not only allocated growth and decline across the landscape, but did so within a clear temporal dynamic.

Economic restructuring, additionally, is manifested at a variety of spatial scales (Smith, 1984: 135-147). Peet (1986), in focusing on the restructuring of the manufacturing sector, situates this transformation in a global context. Beginning in the late 1960s, opposition from labor in core regions, such as the United States and Western Europe, propelled manufacturing operations to Third World countries where opposition was nonexistent and wage levels low. The developed world thus experienced industrial devolution, while a precarious and "disorganic development" took place in peripheral, developing countries. This spatial restructuring, nonetheless, did not dispel the underlying contradictions of manufacturing under capitalism: "Escaping from one (labor) problem at one (national) scale, capitalism enters a new (economic) crisis at a wider (global) scale" (Peet, 1986: 94).

Sassen (Chapter 3) sketches the uneven development of New York as a global city. Neighborhoods of high affluence for the workers in the corporate-administrative-financial headquarters are interspersed with low-wage, immigrant neighborhoods and neighborhoods of black and Hispanic poor. The downsizing of manufacturing and the concentration of producer services and control functions in the city increased the relative demand for and the affluence of professional-managerial labor, thereby encouraging gentrification (another instance of uneven spatial development) and the clustering of these workers in Manhattan. Both formal and informal economies expanded to service these households, with the latter occupied mainly by immigrants. While portions of the informal economy located close to affluent neighborhoods, others sprouted in the outer boroughs. Low-income black and Hispanic communities were displaced by new development and relocated into more peripheral neighborhoods with only tenuous links—often through the provision of menial services—to the "core" economy. Thus economic restructuring affected the pattern of investment and the distribution of population across and within boroughs; that is, at different spatial scales. Space was redifferentiated to reflect the needs of a restructured city.

Spatial differentiation, however, is not uneven development. Many of the spatial differences in activities, investment, and value might simply be due to agglomeration effects and historical patternings, rather than the result of the actual interactions of places of different value and use. That is, uneven spatial development goes beyond a recognition of geographic unevenness to articulate the mutually supportive and generative relationship between places having uneven levels of development. More precisely, capitalism restructures in time and space in ways that link places of growth with places of decline, developed countries with underdeveloped countries, boomtowns with cities in distress.

> Capitalism attempts to move geographically in such a way that it continually exploits the opportunities of development without suffering these economic costs of underdevelopment. That is, capital attempts to see-saw from a developed to an undeveloped area, then at a later point back to the first area which is now underdeveloped, and so forth [Smith, 1984: 49].

Houston's development of its port in the early twentieth century helped to produce Galveston's decline (Platt, 1983), and Houston's early growth in the petrochemical industry was tied to the growth of automobile assembly in the Detroit region (Hill and Feagin, 1987). The economic resurgence and recent international economic success of Japan is part of the economic restructuring in the United States that has seen major industries (e.g., automobiles) crippled and others (e.g., home electronics, motorcycles) virtually eliminated. Silicon Valley's prosperity is based on the employment of low-wage production workers segregated residentially in the city of San Jose, while high-wage technicians, managers, and scientists live in affluent suburbs. San Jose suffers from congestion and environmental pollution, inadequate housing, and a weak fiscal base; but the suburbs are without such problems (Saxenian, 1984).

The linkages of one place to another and their transformation over time to produce uneven development are being articulated within the literature on economic restructuring within the United States. Scholars of international development, however, are more sensitive and precise (e.g., Peet, 1986), while those who study regions (see Sternlieb and Hughes, 1978) or metropolitan areas (see Fainstein et al., 1986) within this country document the consequences of those linkages—for example, migrations of people and factories—but do not specify how capital utilizes uneven development at higher spatial scales to drive capital accumulation.[8] One exception is the work of Allen Scott (1982), who argues for greater attention to the existence of production complexes integrating industries across and within spatial scales, thus providing a sense of the channels along which capital exercises its power to implement and reproduce uneven development. Another exception is David Perry's (1987) analysis of dependent deindustrialization in the Buffalo, New York, region, a restructuring that has not yet included a shift to a service-based economy and that began with absentee ownership and a consolidation of manufacturing firms in the first three decades of the century.

This notion of spatial scales, in turn, is reminiscent of the base-superstructure metaphor. Both divide the world into levels. Spatial scales are nested: the urban within the regional, the regional within the nation-state, the nation-state within the global, for example. The base-superstructure metaphor contains a parallel formulation: the economy subsumes civil society, which contains the state, social relations, and culture, and these levels arch over ideology. The point is not that spatial scales are equivalent to structural levels. Instead, the intent is to raise an issue central to economic restructuring and related to the use of spatial scales and structural formulations—the composition of economic restructuring.

Simply stated, is economic restructuring a global or national phenomenon imposed upon places, or, is it a multitude of place-specific phenomena whose aggregative qualities produce national and global restructuring? In essence, the issue is one of emergent properties. The obvious answer is to point out that capitalism is a world system (Chase-Dunn, 1984) with a logic independent of the elements that constitute it. That is, its properties transcend spatial scales. Yet, this response does not gibe with our common sense that places are unique and that activities occur locally even if carried out simultaneously in different regions or nations. Thus no easy answer exists.

As uneven spatial development unfolds across spatial scales, it also fragments the timing of economic restructuring. While international and national forces are undermining the country's manufacturing sector, the uneven distribution of manufacturing activity across the landscape and the linkages within manufacturing and between manufacturing and other industries are patterning the place-specific, temporal experience of economic restructuring. Foreign competition and domestic managerial indifference undermine the demand for American automobiles. Work shifts shorten and layoffs occur in Detroit. Automobile assembly plants cut back on orders for headlights and deplete inventories. Orders for these and a variety of intermediate inputs are canceled and plants in Anderson, Indiana, face production line shutdowns and layoffs. The rubber industry in Akron also experiences the ripple effect, and it reduces its demand for petrochemicals from Houston. The decline of the steel industry devastates Youngstown and the Pittsburgh region's valley communities, but is hardly felt in Boston where the economy thrives on high-tech, health care, and financial activities. Michigan experiences industrial restructuring before the tire factory towns of Ohio. Youngstown experiences it, and Portland, Maine, does not.

Economic restructuring thus appears as both a temporal and a spatial phenomenon, neither dimension able to stand alone.

Neoclassical promises of market-driven equilibrium, which would obviate temporal unevenness, never materialize. The spatial diffusion of innovation and response is overwhelmed by the inherent seesawing of capitalism between places, never creating a balance; imbalance is what establishes the conditions for profitable investments now and in the future. "The historical rhythm of investment . . . forges specific geographical patterns which in turn strongly influence the agenda of capital accumulation" (Smith, 1984: 125). Thus to ask of economic restructuring when it occurred and whether it took the form of a qualitative discontinuity or a gradual transition is to enter a theoretical thicket whose disentanglement is not certain. Restructuring might well be qualitative and abrupt, yet be so fragmented in its manifestation across space, among structural spheres, and through time, that it can hardly be recognized as a sudden and pervasive novelty. On the other hand, a gradual transition might well be built up with numerous countervailing qualitative changes at various times and in various places. Certainly, resolution of this issue is not impossible, but it is not going to be achieved independently of a sense of the temporal and spatial uneven development of capitalism, and the ways in which capitalism's contradictions are produced and managed.

POLITICAL IMPLICATIONS

Such rethinking of industrial and economic restructuring is not only useful for probing the history and dynamics of capitalism, and thus improving upon our explanations, but it also has political implications as well. Two deserve mention here. First, one's position on the timing and determinants of economic restructuring is central to a political analysis of the perpetuation or eventual demise of capitalism. To view the current instance of restructuring as part of an intrinsic tendency to crisis is to accept the eventual replacement of capitalism by another (usually socialist) social formation. On the other hand, to view the current restructuring as a gradual transition in which capitalism makes continual adjustments in order to reproduce its dominance is ostensibly to adopt a belief that capitalism is not just adaptive but evolutionary in its ability to continue indefinitely.[9]

N. D. Kondratiev, in proposing his theory of long cycles in the 1920s, was attacked by Trotsky for failing to recognize the historical disjunctures of capitalism. Kondratiev was accused of positing internal regulators that maintain capitalism's existence (Day, 1976). By dismissing qualitative discontinuities, Kondratiev became swept up in a political debate focusing on capitalism's eventual demise. Recurring cycles, no matter how long, could not be reconciled with a political commitment to socialism as the eventual destination to be arrived at through capitalist crises. Any interpretation of economic restructuring, as this example illustrates, not only produces a theoretical position but also creates, even if implicitly, an ideological and political one as well.

Theorizing restructuring in both time and space has a second political implication. Only by understanding the processes of restructuring as they occur in different spheres and different places can we understand how conditions for political mobilization are created and begin to consider strategies for resisting economic restructuring (see McDowell and Massey, 1984). Certainly, economic restructuring is not undertaken by labor to improve the lives of workers. Workers' influence over investment decisions and the global rearrangement of production and accumulation must be created through class struggle. Economic restructuring is meant to reentrench capital's interests, at the expense not only of other capitalists, and the state, but at the expense of labor. Given the interdependencies and contingencies, this is a delicate balancing act. Nonetheless, that balance will not tilt to labor independently or in the absence of labor's active resistance to and influence upon economic restructuring. That resistance must begin with a conscious recognition of the specific conjunctures that face labor. It is to this understanding of political resistance that we now turn.

POLITICAL RESPONSE

Despite what seems to be intense restructuring of the economy, with significant reverberations throughout the United States, the late 1970s and the decade of the 1980s have not achieved notoriety for the depth and scope of political unrest. Compared to the diverse and militant social movements of the 1960s and early 1970s and, more appropriately as a comparison, the large-scale political mobilization during the 1930s Depression (Piven and Cloward, 1977), the most recent 10 to 15 years seem quiescent.

Reaction against the rash of plant closings, particularly in unionized industries and where plants were dominant local employers (e.g., Youngstown), has been one exception, but an exception no longer as pronounced. In another sphere, the neighborhood movement has gained strength throughout the last two decades and continues to be a focal point of political organization. Taxpayer revolts in the late 1970s, specifically in California and Massachusetts, seemed to portend the rise of antigovernment fervor, particularly when seen in relation to the rhetoric of Presidents Carter and Reagan. Like the political activity around plant closings and capital mobility, however, that too waned. Feminist organizations, after the defeat of the Equal Rights Amendment, directed their energies to electoral politics. Civil rights groups have traded mass mobilization for local economic development. While gay/lesbian and other interest groups (e.g., antiapartheid organizations on university campuses) periodically mobilize to gain nationwide attention, the norm of relatively unmilitant political activity is sustained (Fainstein, 1987).

Yet, the lack of militancy and collective efforts certainly should not be seen as a lack of political reaction.[10] That reaction has taken a particular form: fragmented along numerous social cleavages, emergent in a variety of social realms, and conservative in political content. It is a renewed anti-immigration sentiment; a reemergence of racial tension involving not just blacks and Hispanics but newly arrived Asians as well; claims of reverse discrimination; a call for trade protection joined with a "Buy American" mentality; the ascendancy of the radical Right; a reinvigoration of Cold War rhetoric and its anti-Soviet propaganda; and the rise of a rigid moralism to be applied to political candidates, a moralism that attempts to eliminate from electoral politics anyone who might have even trifled with countercultural or New Left activities. These are all political reactions to restructuring, but of an individualistic rather than collective orientation.

Such apparent dissonance requires our attention. Why, given that the economy is in such flux and disruption so pervasive, have so few people risen to protest and political reaction taken such a fragmented and conservative form? The answer is deceptively simple: Restructuring is responsible for generating the conditions that make response necessary, but it is also responsible for undermining the conditions that make collective political action possible. Conversely, restructuring encourages political reaction along nationalistic, moralistic, and socially divisive paths. The present economic restructuring of capitalism in the United States, a restructuring whose outcomes—if successful—would reinvigorate capital accumulation and strengthen capital's control of the sphere of production and the state, is taking place at the expense of the collective interests of labor. The cost to labor is not only a loss of influence at the workplace but also a loss of influence over the state. At the same time, the spatial and temporal unevenness of restructuring undermines the bases for collective consciousness and mass organization in response to such threats to labor. Thus two issues concern us in this section: first, the extent to which restructuring has and is occurring in ways that erode the working and living conditions of labor, and, second, the ways in which that restructuring contributes to the present political response.

RESTRUCTURING AND LABOR

As a result of the recent economic restructuring, the position of labor has been eroded at the workplace, in the sphere of consumption, and in its relation to the state. At the workplace, labor has lost influence as the more heavily unionized industries (e.g., steel, automobiles) have responded to foreign competition and internal forces by closing plants and shifting operations outside of unionized areas or to offshore facilities. Such restructuring has resulted in large-scale layoffs and created the leverage for capital to negotiate contracts that require give-backs of benefits, of worker control over the labor process, and even of wages (Bluestone and Harrison, 1982; Clark, 1986a). Moreover, the cutbacks in employment have not only functioned as threats to labor demands for higher wages and greater benefits but also have eroded unions' ability to increase and maintain membership and thus to organize new plants and mobilize for contract negotiations (Clark, 1987). Simultaneously, the movement of employment opportunities out of large-scale manufacturing operations suitable for industrial unions and into smaller-scale service operations has also weakened union influence despite the emergence of white-collar unions.

While the economy has continued to function as a "job machine" (Bluestone and Harrison, 1986), generating increasing amounts of employment opportunities even in times of cyclical economic contractions, that growth hides a significant restructuring of the labor market (Briggs, 1987).[11] First of all, with the shift of employment away from manufacturing and into a range of services, the relatively high-paying jobs of unionized machine operators and assembly-line workers have been replaced by nonunionized, low-wage jobs for restaurant helpers, hospital attendants, and word processors. Contributing to employment loss in key manufacturing industries has been an infatuation with robotics, machines that function as operators and perform the work of multiple workers. In fact, job growth in the 1980s has been concentrated primarily at the low-wage end of the occupational scale. "During the 1970s about one out of every five net additional wage-earners found a job (or jobs) paying as little as $7,000. Since 1979, that fraction has risen to nearly six in ten" (Bluestone and Harrison, 1986: 5).

In addition, the percentage of full-time jobs in the economy has been decreasing as the percentage of involuntary part-time employment has crept up with every recent business cycle (Tokatli, 1987). Moreover, "from 1979 to 1985 nearly 28 percent of all new jobs were part-time" (Levinson and Mishel, 1987: 4). The shift to the service sector, traditionally a heavy user of part-time workers, has coupled with a movement away from full-time labor enjoying a wide range of benefits to produce jobs whose aggregate benefits are less than what would be the case if full-time work was being offered. Thus even while total employment is expanding, the total wage and nonwage benefits obtained through employment are in relative decline. Moreover, the growth in employment is not maintaining pace with the expansion of the labor force itself, thus contributing to higher and higher levels of unemployment despite declining unemployment rates during 1987.

The end results of this restructuring of the labor market have been an erosion of employment opportunities for male household heads and the further polarization of the income distribution. The expansion of service sector and of part-time jobs has benefited mainly those groups with the least social power and the least political power at the workplace—women and minorities. The male labor force participation rate has dropped simultaneously with a rejection of the family wage, an essential element in the postwar contract between capital and labor. As a consequence, more and more people have to work in order to support a family than in the past. No longer can a single male, particularly one in a blue-collar occupation, hope to earn enough to support a family without the assistance of other family members.

Income polarization has been exacerbated by economic restructuring. Those at the low end of the income distribution are earning less, those at the upper end more, and the middle is being squeezed. Between 1970 and 1980, those workers in the top third of the earnings distribution increased from 32.2% to 37.0% of all workers, while those in the bottom third experienced an increment of 6.5% on top of an initial 32.0% (Sassen-Koob, 1985: 261). The middle third of the earnings distribution, however, declined by 32%.[12] The obvious consequence is to make life more difficult for the lower and working classes by forcing them into informal activities to supplement their declining incomes (and thus subtracting from leisure time) and by reducing their level of consumption. At the same time, the middle earning classes are also losing ground, thus exacerbating class polarization. In a society in which food costs seldom decline, transportation costs continually increase, and housing is increasingly unaffordable, one must earn more and more merely to maintain the status quo. For most of the laboring class, this is not happening.

Labor is not simply losing influence and employment opportunities at the workplace, and finding consumption and reproduction more and more difficult; labor is also losing influence over the state and in the political sphere where labor might establish arguments that protect and enhance its position at the workplace. Certainly unions have come under attack as national, state, and local governments have touted "good business climates" (a euphemism for right-to-work states and nonunionized labor forces) as the key to the attraction of investment and thus to economic growth. In a parallel fashion, the conservative program of the Reagan administration infused the National Labor Relations Board with an ideology antagonistic to unions (Clark, 1986b).

More basically, the continual struggle for control over the state by capital and labor has clearly swung toward capital. Not only have the last two federal administrations withdrawn from the notion of a strong federal government protecting labor and enhancing human welfare, but governmental subservience to capital investment has deepened across a variety of policy areas. Privatization and public-private partnerships are key thematic elements. One sees it very explicitly in local property development, but it is also evident in the social services, foreign trade, and technological research. Additionally, while progressive politicians have emerged to champion a reform agenda (Clavel, 1986), elected officials in general are reluctant to oppose the interests of capital and quick to associate themselves with its interests (Kantor, 1987).

Reasons to engage in political resistance to economic restructuring, then, are easy to find. Labor is losing resources, employment opportunities, and political influence; and the consequences are making life harsher. Moreover, the employment opportunities that do exist are increasingly undesirable and inadequate. The conditions for class consciousness and political mobilization are present. Most important, life has become worse for many factions of labor at a time when other factions are doing extremely well and when the national media are generally bullish on the state of the economy and on the new, postindustrial society of professionals and educated labor. Admittedly, this is not a time of collective suffering, nor are we experiencing an unquestionable and precipitous erosion of economic prosperity. Rather, it is a period of relative deprivation within the ranks of labor, between labor and capital, and, for certain factions of labor, between the past and the present. "The restructured capitalist economy, having subjected labor to the disciplines of severe recession and depressed wage rates, is providing new structural conditions for movements uniting previously disparate portions of the population" (Fainstein, 1987: 326). Why, then, the inward turn of political response and the quiescence, passivity, and resignation?

UNEVEN DEVELOPMENT AND POLITICAL ACTION

The key to this conundrum lies in the spatial and temporal unevenness of economic restructuring. Restructuring's dynamics fragment its negative impacts, undermine the conditions for collective consciousness and mass mobilization, and generate obstacles to political action. "The unfolding of capital restructuring creates profoundly destabilizing conditions for everyday life" (Feagin and Smith, 1987: 24), and challenges the political response that does emerge. Capital accomplishes this by intermixing growth and decline in close spatial proximity, thus creating cross-cutting cleavages within labor, and by developing a specialized space-economy that restructures industries at different times and communities with different degrees of severity. These effects are exacerbated by the spatial mobility of labor and capital, allowing labor an alternative to political resistance and capital a compensating effect. In addition, the uneven timing of restructuring, the erosion of dramatic consequences through a multitude of subtle adjustments, and the complexity of complementary and opposing changes outside the sphere of production also contribute to a barren field for the cultivation of grass-roots movements. As a result, capital does not sit as an easy target, and political consciousness is not easily manufactured.

The contemporary restructuring of capital involves neither recession nor prosperity. Rather, what makes it so profound, and such a dilemma for theoreticians and policymakers alike, is its unevenness across industries, communities, segments of labor, neighborhoods, and even firms within industries. Though certain labor markets, particularly those with highly specialized economies—Detroit and Pittsburgh come readily to mind—seem to be overwhelmed by job loss, high unemployment, and recession; and others such as Silicon Valley and Route 128 experience labor shortages; most communities and labor markets have been struck by a combination of growth and decline. Philadelphia's central business district was a hotbed of new construction and job growth in the mid-1980s, while manufacturing and commercial activity in North Philadelphia, only a few miles away, continued to be weak (Beauregard, forthcoming-b). In Los Angeles, the homeless population has expanded, high technology and advanced services have boomed, and immigrant labor has been absorbed into illegal and legal, low-wage apparel shops (Soja et al., 1983). In most instances, restructuring at any spatial scale encompasses prosperity and poverty, new investment and disinvestment, and a reallocation, not a wholesale reduction, of employment opportunities.[13]

The unevenness of the space economy, then, affects the distribution of the costs and benefits of restructuring across space and throughout the labor force. Specialized economies reflect growth or decline, depending on the industries that they happen to contain. More diverse economies present more of a mix of objective conditions. In the former, the source of discontent is often obvious: the dominant industry and the corporations that control it. The discontent is frequently channeled through the unions and into collective bargaining or employee buy-out options, and workers frequently turn to government for assistance (Lauria, 1986). In other communities with dominant industries, however, resignation occurs (Nash, 1987). The conditions are clear, mechanisms for organization available, and responsible parties apparent. Active resistance is more probable but still not preordained.

In communities with a diverse economic base, similar conditions do not exist. Some factions of labor suffer, others maintain, and still others do well—often within the same occupational and industrial categories. Even if unions exist, different unions in dissimilar industries find it difficult to combine forces when the causes and consequences of restructuring are not equivalent. Workers have fewer channels by which to link their complaints and understandings. The government, moreover, finds it easy to avoid involvement by isolating the problem and by pointing to growing industries and expanding employment opportunities for those displaced by progress. The result is often a demoralized labor force unwilling to resist restructuring because it believes that it lacks control and that its efforts will be fruitless (Murgatroyd and Urry, 1985: 51).

On a regional scale, this results in a territorial variation in political responses not easily combined into a common national front against economic restructuring. Markusen's (Chapter 4) description and analysis of regional politics in the United States makes this clear. The blatant pursuit by southern politicians and booster groups of migrating firms from the North not only undermines capital investment in the latter region through its exacerbation of disinvestment, but also further weakens labor in the Rustbelt as unionized jobs are diminished and the benefits of nonunionization and right-to-work legislation are demonstrated to be successful in attracting economic growth. Even within regions, the politics of environmental protection, growth controls, and low-density development that often surround high-tech agglomerations serve well the affluent and mainly suburban communities of venture capitalists and research scientists but are antagonistic to the needs of assembly workers virtually trapped in the older, central city. The political issues related to restructuring are quite distinct, thus undermining any possibility of a common strategic response.

Economic specialization across space additionally affects the depth to which restructuring strikes a region or community. Detroit is ravaged by unemployment and poverty, while Indianapolis, a few hundred miles southeast, experiences only indirect hits and ripple effects. The mill towns of the Pittsburgh region face the choice of becoming "bedroom suburbs" for service workers, or ghost towns. Yet, the central city itself, seemingly dependent upon the productive capacity once located in its hinterland, has found other specializations to buffer many of its citizens from industrial restructuring. Kensington, the center of Philadelphia's textile and apparel industry, becomes a "white ghetto" while Old City, the former warehouse and light manufacturing core of the city, is rediscovered by residential and commercial investors intent on gentrifying the area (Beauregard, forthcoming-b). Thus the nature of the consequences and their intensity are quite different across space, and this cannot but obstruct an awareness of economic restructuring and collective activities to oppose it.

This impact of spatial economic specialization on political resistance suggests an interesting perspective on the prevailing ideology of subnational economic development strategy, particularly its emphasis on industrial diversity. The current wisdom, in great part a response to contemporary restructuring, is that specialized economies, despite rapid growth and the spin-offs of agglomeration, are too susceptible to economic restructuring and must be replaced by more diversified economies. A mix of industries, none of which dominates the labor market or the flow of capital within the region or community, is ideal. If foreign competition threatens or technology makes an industry obsolete, others will continue to prosper and the negative impacts will be concentrated only in a specific (and hopefully small) segment of the labor market.

From a political perspective, diversification serves more than the economic interests of community leaders; it also serves the political interests of capital. Large industrial firms are, all other things being equal, more susceptible to unionization, and this tendency is reinforced when many workers in the community are employed by a single or even many similar firms. Moreover, diversification fragments the timing and impact of restructuring. The awareness of labor's troubles, then, is not obvious to all workers at the same time. Thus, while diversification might protect workers against unemployment, by offering a range of employment opportunities, it also weakens labor politically.[14]

Further enervating political resistance is the permeability of state and municipal boundaries. Labor, as capital, can migrate: Relocation is an alternative to political resistance. The travails of the Rustbelt in the mid-1970s led to a major increase in the flow of labor to the Sunbelt. Not only did these population shifts help to right imbalances in specific labor markets (while often creating imbalances in others), they also removed potential protesters from the community. The spatial mobility of labor thereby serves capital in two ways: economically and politically.

Less recognized, and studied, is the fact that labor can simply shift commuting patterns, because growth and decline are so closely situated in space. In the Pittsburgh region, for example, the rise of a robust service sector in the educational, science, and hospital sectors within the city; the decline of manufacturing jobs in the valley towns; and the rise of office and retail activity along the suburban corridor leading to the airport (Spatter, 1987) could not help but increase and refocus commuting. The political consequence is to realign place identifications and to encourage residential mobility from communities not experiencing growth.

While capital migration is a villain in the restructuring process, capital migration also counteracts the negative consequences that ensue. Capital itself is not a homogeneous, functional entity; and economic decline creates conditions that benefit certain factions of capital (and labor). Areas of high unemployment, weakened unions, high vacancy rates in industrial and commercial buildings (and thus relatively low rents), and inexpensive land and buildings within a weak property market can be attractive to capital. In Houston, for example, investors stopped financing new office construction when falling oil prices led to extremely high vacancy rates in office buildings (Feagin, 1987). Other investors then began to purchase property, gambling that the market would rebound and that they would profit from having bought at bargain-basement prices (Hooper, 1986). Areas of Los Angeles and New York City, or distressed cities such as Newark, New Jersey, become attractive sites for low-wage assembly operations. Unemployment is high, and this makes labor cheap and local governments willing to provide a host of investment subsidies. Labor, moreover, is disorganized. Capital, therefore, can be expected to migrate to areas suffering from restructuring caused by capital out-migration. By doing so, the political situation is further confounded, siphoning off discontent through employment and reinforcing the ties to government through the benefits it brings.

Of course, in some instances, the schizophrenic nature of capitalist investment might open up opportunities for political action. This seems to be the case with the recent attempts to legislate hiring preference ordinances in a number of large cities. As Beauregard (Chapter 5) indicates for Philadelphia, the decline in manufacturing jobs was accompanied in the 1980s by an expansion of construction investment, particularly within the central business district. Minority groups turned to the local government for support in obtaining employment in the construction industry. These jobs would nearly match in wages those formerly provided by manufacturing. Moreover, by making local governments dependent on local property investment, capital (with the help of the state) had generated a new channel for political action, one directed toward capital's prerogatives in the area of employment relations. Mixed results occurred: Minorities were provided with opportunities for political victories but labor was divided and the larger potential for resistance weakened. Nonetheless, "popular praxis matters . . . and when it is present, the balance of power within the state, and hence the state role in urban restructuring, becomes more than a matter of capital accumulation" (Feagin and Smith, 1987).

Uneven spatial development also fractures political resistance through its effect on the timing of economic restructuring. Because restructuring occurs in different places and, within those places, at different intensities, and then diffuses across the landscape, one effect is an uneven temporal pattern. Quite diverse places restructure at clearly distinct times. Large corporations do not close down or reduce production at all of their plants simultaneously. Rather, shutdowns, layoffs, and start-ups are phased. To the extent to which the plants are located in different communities, the economic restructuring that caused the corporate reorganization is experienced at different times by workers. Or, in terms of one of the more theoretically interesting manifestations of economic restructuring, gentrification in a city does not begin in the same year in all neighborhoods that will eventually be gentrified (Beauregard, 1986). A concentration of activity in one or a few neighborhoods occurs and then, when the process has proceeded to a particular point, diffuses into other neighborhoods. Timing is uneven.

This temporal variability of restructuring has a detrimental effect on political resistance. It is difficult for workers in one community to establish organizational ties with those in another who have yet to experience, and who thus hope to avoid, the employment consequences of restructuring. Industrial unions are placed in a precarious position as workers from different local plants pressure national union leaders to take quite dissimilar actions. Even within communities, a shared sense of oppression does not leap readily from those not yet suffering, despite indications that hardship is imminent. Just as space fractures collective consciousness, timing does also. In addition, it makes strategy more complex. Resistance groups must not only mobilize to oppose capital but do so in ways that recognize that their membership consists of individuals with different understandings of restructuring and with clearly diverse interests in what constitutes a resolution of their difficulties.

From an organizing perspective, the ideal situation is a deep and sudden restructuring whose consequences are focused in a single industry, clearly visible, and personally felt. Organizers need restructuring to be dramatic. The nature of this process is frequently, but not always, the opposite. Certainly the present restructuring of capital has presented itself in a number of striking ways: plant closings in steel and automotive assembly, the large federal deficit, the sudden rise in gasoline prices, the trade imbalance of the 1980s, the near collapse of savings and loan associations in the mid-1970s, the major increments to inflation and unemployment in the last decade, and the precipitous and steep stock market decline of late 1987. Not all of these were sudden events, and many of the events were preceded by a multitude of eventually crippling decisions. For the most part, however, restructuring is more subtle, occurring through a variety of relatively small (but no less significant) adjustments to the labor process, investment patterns, and productive output (Clark et al., 1986). Such adjustments might or might not be perceived by labor (or even capital) as part of a larger structural shift in the economy.

The actions of capital and the consequences for labor are also complex in another way. Large-scale changes and incremental adjustments occur in the sphere of production through a melange of decisions. Often labor does respond: protesting the reassignment of tasks, resisting wage reductions, shifting consumption patterns, and relocating. Capitalism is not a one-sided affair. Labor and capital struggle over the nature of society. The playing field, however, is tilted in favor of capital: It controls the means of production and the bulk of resources in society. As a result, capital has the initiative and the edge, even if antagonisms within capital's camp weaken its efforts. For labor, the initiative is not so easily taken. Moreover, the machinations of capital are not always perceived as they happen but only through consequences experienced later.

Thus labor finds itself in a reactive position. Combine that with the lack of knowledge about how changes in the economy reverberate throughout the various spheres of society, and the problems of defining a collective consciousness and organizing increase exponentially. Most obvious are the manifestations of restructuring at the workplace. Labor, however, does not live only at the workplace, and some people do not even enter that realm. Certainly economic restructuring reaches throughout society, but neither as the sole driving force of all changes nor the clear cause of the remainder. Economic restructuring is refracted once again, this time by the complexity of the social formation itself.

As suggested above, labor has always had access to the state, and that access can become leverage over capital. In the quintessential company town, of course, capital and the state are so interdependent as to thwart labor at every turn. In communities where the local government is relatively autonomous from capital, labor can often gain influence, if not directly then through federal and state channels. At times, labor has even been able to gain control of the state from capital (Clavel, 1986) and to redirect state actions mainly toward labor's interests. Giloth and Mier (Chapter 6) describe Chicago under Mayor Harold Washington during the 1980s when a neighborhood-focused and employment-centered (rather than capital-centered) economic development policy was established. Because economic restructuring in Chicago weakened the cohesiveness of capital and made life difficult for minorities and white working-class groups alike, labor was able to expand and institutionalize its power. Sadly, these are exceptions and not the norm.

In sum, while the present economic restructuring is proceeding in ways that impose net costs upon labor, that restructuring is also occurring in a fashion that erodes collective consciousness, mass mobilization, and political action. The uneven spatial development of capitalist society, and the concomitant and resultant temporal fracturing of restructuring, create a complex mosaic of consequences that mix growth and decline, occur in different communities at different times, penetrate unevenly across factions of labor, and manifest themselves in quite diverse ways within the various spheres of society. Thus, while the evidence exists to allow academics and other observers to proclaim the economic restructuring of capital, that knowledge is neither of the type nor intensity—much less broadly shared—that labor's collective political resistance is easily forthcoming.

CONCLUSIONS

Economic restructuring is a profoundly spatial and temporal phenomenon. New relations and their consequences have spatial extension, and new processes occur in different places in different ways at different times. Neither the crises of capitalism nor its present restructuring can be viewed independently of the ways in which space and time matter. Economic restructuring is not just a locational rearrangement of product mix and the social relations of production. Space plays an active role in its own transformation, and uneven spatial development is central to our understanding of economic restructuring.[15] In fact, the leverage that capital holds over labor, not only during qualitative restructuring but also during relatively quiescent periods and even crises, emanates from prior waves of uneven development.

While unique locations are confronted with different intensities of restructuring, uneven spatial development additionally affects the timing of restructuring. That timing, however, is inseparable from the spatial unevenness of restructuring—it cannot be perceived independently of its place-specific manifestation. Restructuring is about where capital, and thus labor, will function, and how space can be used to enhance capital accumulation and forestall systemic crisis. Our judgment of the meaning and importance of economic restructuring is embedded in a spatial and temporal context that refracts it, thereby hindering abstraction and impeding comprehension and theorization.

At the same time, while it is wrong to view economic restructuring outside of spatial and temporal realms, it is also wrong to accept restructuring as wholly an economic phenomenon. The more precise phrasing would refer to the restructuring of the political economy.[16] Restructuring entails attempts by capital to reassert its dominance over labor at the workplace, to penetrate further the state and involve it more deeply in the accumulation process, and to weaken the ties between labor and the state. The relations among these entities are both economic and political; the one reinforces the other. Economic restructuring cannot proceed otherwise under capitalism. The separation of the economy from civil society is only ideologically resolved. A capitalist state sits at the center of civil society. Moreover, the dominant ideology, adopted by academics and reinforced by their disciplinary boundaries, deflects attention from the inherently political nature of the economy. The power of capital is not only predicated on a state that buffers capital from class struggle and protects capital from its own greed, but also upon labor's belief that politics are something to engage in outside the workplace (Katznelson, 1981).

Thus, as economic restructuring unfolds, capitalism reestablishes its political strength. It does so on the "backs of labor" (or at least certain factions of labor) and at the expense of factions of capital both within and outside the country. For just as restructuring exhibits itself and operates spatially in the sphere of production, it also has spatial extension throughout the political realm. Most important, the uneven spatial and temporal development of restructuring undermines the conditions for collective consciousness and mass mobilization, thus enabling capital to proceed with minimal resistance. In this way, the intrinsic spatiality of economic restructuring and its undeniable political dimension are linked.

Theoretical indecisiveness on issues of incidence, timing, depth, and process of economic restructuring, then, can be traced to the complexity and indeterminacy generated by the uneven spatial and historical development of capitalism. Restructuring, metaphorically speaking, is not a stone dropped into a still pond. As the consequences of economic restructuring ripple through the various spheres of society and reverberate back to the economy, they take circuitous and intersecting paths. There is structure, but the contingency of its constant production and reproduction make observation difficult and theorization problematic. What is irrefutable is that the study of economic restructuring cannot occur unmindful of space and time, and in ignorance of its political impetus and manifestations.

NOTES

1. Peet (1987: 2), for example, uses *industrial restructuring* to refer to what I call "economic restructuring."

2. For some theorists, *economic restructuring* and *urban restructuring* are equivalent. Soja et al., (1983: 195), for example, state: "A comprehensive process of *urban* restructuring has been taking place since the late 1960s, affecting the organization of the

labor process and the composition of the work-force, the location of industry and the sectoral distribution of employment, the organization of the working class and the patterns of class conflict."

3. The statement corresponds to Joseph A. Schumpeter's much-cited notion of creative destruction: the incessant revolutionizing of capitalism from within through constant creation and destruction (Schumpeter, 1942: 81-86).

4. Williams (1977: 75) writes that "the proposition of a determining base and a determined superstructure" is key to Marxist analysis but is metaphorical rather than precisely conceptual (p. 77).

5. Weiss (1987), for example, discusses how subdivision developers and builders in the 1920s utilized the state to rationalize chaotic land and housing markets.

6. Giddens (1979) is certainly one theorist who appreciates this point and has proposed solutions.

7. Marris (1987: 120) comments in regard to this point that "any interpretive scheme which attempts to organize the flow of events into a sequence of historical periods is open to the objection that the crucial transformations began long before a period is said to end and continue long after its successor is said to have begun."

8. A fruitful path of inquiry here might be the process of industrial subcontracting (see Holmes, 1986).

9. Chase-Dunn (1984) seems to represent this latter political stance, while Bradbury (1985) and Castells (1980) hold to the former.

10. I owe this very important point to Bob Lake, who also provided other substantive and useful criticisms of an earlier draft.

11. From 1976 to 1980, total employment grew by 7.3 million, and "since 1981 the total number of jobs in the U.S. has expanded by nearly 10 million, despite the severe 1981-82 recession" (Bluestone and Harrison, 1986: 2).

12. Levinson and Mishel (1987: 5) found similar shifts in family income between 1973 and 1985. The category earning below $20,000 per year expanded 3.3%, that earning between $20,000 and $50,000 diminished by 5.1%, and the top category earning above $50,000 increased by 1.8%.

13. Perry (1987: 113) argues that this is not the case for the Buffalo region: "Buffalo and the Western New York region are caught up in a downward spiral of deindustrialization—neither 'restructured' to compete as a prominent participant in the new round of technologically driven advances in information and communications, nor 'restructured' to compete as a leader in the sectoral shift of the United States into service production."

14. Nash (1987: 295) presents a twist on this argument. She claims that the disinvestment of the General Electric Corporation from Pittsfield, Massachusetts, where it had been the dominant employer, has resulted in the erosion of a set of values (e.g., individualism, competitiveness) that underpinned the corporation's influence in the community. With restructuring, "management is destroying the basis for their [sic] own hegemonic control."

15. Only hinted at in this chapter is capital's active use of space to facilitate restructuring (see Massey, 1984).

16. I think this is what is meant by those who use the term *urban restructuring* (e.g., Feagin and Smith, 1987; Soja et al., 1983).

REFERENCES

BEAUREGARD, R. A. (1986) "The chaos and complexity of gentrification," in N. Smith and P. Williams (eds.) Gentrification of the City. Boston: Allen & Unwin.

BEAUREGARD, R. A. (1988) "History and dynamics in regional adjustment theory." Urban Geography 9 (2): 302-310.

BEAUREGARD, R. A. [ed.] (forthcoming-a) Atop the Urban Hierarchy. Totowa, NJ: Rowman & Littlefield.

BEAUREGARD, R.A. (forthcoming-b) "The spatial transformation of postwar Philadelphia," in R. A. Beauregard (ed.) Atop the Urban Hierarchy. Totowa, NJ: Rowman & Littlefield.

BLUESTONE, B. and B. HARRISON (1982) The Deindustrialization of America. New York: Basic Books.

BLUESTONE, B. and B. HARRISON (1986) "The great American job machine." Study prepared for the Joint Economic Committee, U.S. Congress, Washington, DC.

BLUESTONE, B. and B. HARRISON (1987) "The grim truth about the job miracle." New York Times (February 1).

BOYER, M. C. (1983) Dreaming the Rational City. Cambridge: MIT Press.

BRADBURY, J. H. (1985) "Regional and industrial restructuring processes in the new international division of labor." Progress in Human Geography 9 (1): 38-63.

BRIGGS, V. M., Jr. (1987) "The growth and composition of the U.S. labor force." Science 238: 176-180.

CASTELLS, M. (1980) The Economic Crisis and American Society. Princeton, NJ: Princeton University Press.

CHASE-DUNN, C. K. (1984) "The world-system since 1950: what has really changed?" in C. Bergquist (ed.) Labor in the Capitalist World-Economy. Beverly Hills, CA: Sage.

CLARK, G. L. (1986a) "The crisis of the midwest auto industry," in A. J. Scott and M. Storper (eds.) Production, Work, Territory. Boston: Allen & Unwin.

CLARK, G. L. (1986b) "Restructuring the U.S. economy: the NLRB, the Saturn project, and economic justice." Economic Geography 62 (4): 289-306.

CLARK, G. L. (1987) "The geography of US union elections 1." Environment and Planning A 19: 33-57.

CLARK, G. L., M. GERTLER, and J. WHITEMAN (1986) Regional Dynamics. Boston: Allen & Unwin.

CLAVEL, P. 1986. The Progressive City. New Brunswick, NJ: Rutgers University Press.

DAY, R. B. (1976) "The theory of the long cycle: Kondratiev, Trotsky, Mandel." New Left Review 99: 67-82.

FAINSTEIN, S. S. (1987) "Local mobilization and economic discontent," in M. P. Smith and J. R. Feagin (eds.) The Capitalist City. Oxford: Basil Blackwell.

FAINSTEIN, S. S. et al. (1986) Restructuring the City. New York: Longman.

FEAGIN, J. R. (1987) "The secondary circuit of capital: construction in Houston, Texas." International Journal of Urban and Regional Research 11 (2): 172-192.

FEAGIN, J. R. and M. P. SMITH (1987) "Cities and the new international division of labor: an overview," in M. P. Smith and J. R. Feagin (eds.) The Capitalist City. London: Basil Blackwell.

GIDDENS, A. (1979) Central Problems in Social Theory. Berkeley: University of California Press.

GLICKMAN, N. J. (1987) "Cities and the international division of labor," in M. P. Smith and J. R. Feagin (eds.) The Capitalist City. Oxford: Basil Blackwell.

HILL, R. C. and J. R. FEAGIN (1987) "Detroit and Houston: two cities in global perspective," in M. P. Smith and J. R. Feagin (eds.) The Capitalist City. Oxford: Basil Blackwell.
HOLMES, J. (1986) "The organization and locational structure of production subcontracting," in A. J. Scott and M. Storper (eds.) Production, Work, Territory. Boston: Allen & Unwin.
HOOPER, C. (1986) "City review: Houston." National Real Estate Investor (June): 161-175, 218-222.
JAMESON, F. (1984) "Postmodernism, or the cultural logic of capitalism." New Left Review 146: 53-92.
KANTOR, P. (1987) "The dependent city: the changing political economy of urban economic development in the United States." Urban Affairs Quarterly 22 (4): 493-520.
KATZNELSON, I. (1981) City Trenches. Chicago: University of Chicago Press.
KAY, G. (1979) The Economic Theory of the Working Class. New York: St. Martin.
LAURIA, M. (1986) "Towards a specification of the local state." Antipode 18 (1): 39-63.
LEVINSON, M. and L. MISHEL (1987) "Whose recovery is it, anyway?" Democratic Left 15 (4): 4-6.
MARKUSEN, A. R. (1985) Profit Cycles, Oligopoly and Regional Development. Cambridge: MIT Press.
MARRIS, P. (1987) Meaning and Action. London: Routledge & Kegan Paul.
MASSEY, D. (1984) Spatial Divisions of Labor. New York: Methuen.
MASSEY, D. and R. MEEGAN. (1982) The Anatomy of Job Loss. New York: Methuen.
McDOWELL, L. and D. MASSEY (1984) "A women's place?" in D. Massey and J. Allen (eds.) Geography Matters! Cambridge: Cambridge University Press.
MOOS, A. I. and M. J. DEAR (1986) "Structuration theory in urban analysis: 1 theoretical exegesis." Environment and Planning A 18: 231-252.
MURGATROYD, L. and J. URRY (1985) "The class and gender restructuring of the Lancaster economy, 1950-1980," in L. Murgatroyd et al. (eds.) Localities, Class, and Gender. London: Pion.
NASH, J. (1987) "Community and corporations in the restructuring of industry," in M. P. Smith and J. R. Feagin (eds.) The Capitalist City. London: Basil Blackwell.
PEET, R. (1986) "Industrial devolution and the crisis of international capitalism." Antipode 18 (1): 78-95.
PEET, R. (1987) "Introduction," in R. Peet (ed.) International Capitalism and Industrial Restructuring. Boston: Allen & Unwin.
PERRY, D. (1987) "The politics of dependency in deindustrializing America," in M. P. Smith and J. R. Feagin (eds.) The Capitalist City. London: Basil Blackwell.
PIVEN, F. F. and R. A. CLOWARD (1977) Poor People's Movements. New York: Pantheon.
PLATT, H. L. (1983) City Building in the New South. Philadelphia: Temple University Press.
SASSEN-KOOB, S. (1985) "Capital mobility and labor migration: their expression in core cities," in M. Timberlake (ed.) Urbanization in the World-Economy. New York: Academic Press.
SAXENIAN, A. (1984) "The urban contradictions of Silicon Valley," pp. 163-197 in L. Sawers and W. K. Tabb (eds.) Sunbelt/Snowbelt: Urban Development and Regional Restructuring. New York: Oxford University Press.

SCHUMPETER, J. A. (1942) Capitalism, Socialism and Democracy. New York: Harper & Row.

SCOTT, A. J. (1982) "Production system dynamics and metropolitan development." Annals of the Association of American Geographers 72 (2): 185-200.

SMITH, N. (1984) Uneven Development. London: Basil Blackwell.

SOJA, E. W. (1987) "Economic restructuring and the internationalization of the Los Angeles region," in M. P. Smith and J. R. Feagin (eds.) The Capitalist City. London: Basil Blackwell.

SOJA, E. W., R. MORALES, and G. WOLFF (1983) "Urban restructuring: an analysis of social and spatial change in Los Angeles." Economic Geography 59 (2): 195-230.

SPATTER, S. L. (1987) "City review: Pittsburgh." National Real Estate Investor 29 (11): 275-280, 304-308.

STERNLIEB, G. and J. W. HUGHES (1978) Revitalizing the Northeast. New Brunswick, NJ: Center for Urban Policy Research.

STORPER, M. and A. J. SCOTT (1986) "Production, work, territory: contemporary realities and theoretical tasks," in A. J. Scott and M. Storper (eds.) Production, Work, Territory. Boston: Allen & Unwin.

TOKATLI, N. (1987) "The rise of involuntary part-time employment in the U.S. labor market, 1963-1985." Unpublished paper [available from author at Department of Urban Planning, Rutgers University, New Brunswick, NJ 08903].

WEISS, M. (1987) The Rise of the Community Builders. Berkeley: University of California Press.

WILLIAMS, R. (1977) Marxism and Literature. New York: Oxford University Press.

WILLIAMS, W. A. (1973) The Contours of American History. New York: New Viewpoints.

Index

About the Contributors

ROBERT A. BEAUREGARD is Associate Professor of Urban Planning and Policy Development at Rutgers University. His work focuses on urban economic development and labor markets, urban development, and redevelopment with particular emphasis on gentrification and the city building process, and planning theory. He is coauthor of *Revitalizing Cities* (Association of American Geographers, 1981) and editor and contributor to *Atop the Urban Hierarchy* (Rowman & Littlefield, forthcoming).

GORDON L. CLARK is Professor of Labor Studies and Urban Policy in the School of Urban and Public Affairs at Carnegie Mellon University. He most recently coauthored *Regional Dynamics: Studies in Adjustment Theory* (Allen & Unwin, 1986) and *State Apparatus: Structures and Language of Legitimacy* (Allen & Unwin, 1984) and is author of *Judges and the Cities* (University of Chicago Press, 1985) and *Inter-Regional Migration, National Policy and Social Justice* (Rowman & Littlefield, 1983).

NORMAN I. FAINSTEIN is Dean of Liberal Arts and Science and Professor of Sociology at Baruch College of the City University of New York. He has written extensively on urban political economy and policy, and is coauthor of *Restructuring the City* (Longman, revised edition, 1986).

SUSAN S. FAINSTEIN is Professor of Urban Planning and Policy Development at Rutgers University. She is author of numerous books and articles on urban political economy and urban redevelopment, and coauthor of *Restructuring the City* (Longman, revised edition, 1986).

ROBERT P. GILOTH is a Ph.D. candidate in the Department of City and Regional Planning at Cornell University and former Deputy

Commissioner of Economic Development for the City of Chicago. Prior to that appointment he was active in community and economic development in Chicago, including tenure as executive director of the 18th Street Development Corporation and founding board member of the Chicago Rehab Network.

ANN ROELL MARKUSEN is Professor of Business and Social Policy at Northwestern University. Her teaching and research concern interregional growth disparities and urban and regional policy, regions and regionalism, state and local economic development planning, analysis and planning for women, and metropolitan public finance. Her books include *The Politics of Regions* (Rowman & Littlefield, 1987), *Profit Cycles, Oligopoly, and Regional Development* (MIT Press, 1985), and *Silicon Landscapes* (Allen and Unwin, 1985) with Peter Hall.

ROBERT MIER is currently Deputy Mayor of the City of Chicago. Previous to that he was the city's Commissioner of Economic Development, and Director of the Center for Urban Economic Development at the University of Illinois at Chicago Circle. He has written and published widely on a variety of economic development and labor market themes: high technology, Hispanic employment opportunities, nonprofit business activities, enterprise zones, subemployment, and antipoverty strategies.

SASKIA SASSEN is Associate Professor and Chair of the Program of Urban Planning in the Graduate School of Architecture and Planning at Columbia University. Her work focuses on immigration and informalization in global cities, specifically New York and Los Angeles. She is the author of *The Mobility of Labor and Capital* (Cambridge University Press, 1987).

NOTES

NOTES